Legion of the Lancasters

Legion of the Lancasters

Martin W. Bowman

Pen & Sword
AVIATION

First published in Great Britain in 2022 by
Pen & Sword Aviation
An imprint of
Pen & Sword Books Ltd
Yorkshire – Philadelphia

Copyright © Martin W. Bowman 2022

ISBN 978 1 52674 607 8

The right of Martin W. Bowman to be identified as Author of this work has been asserted by him in accordance with the Copyright, Designs and Patents Act 1988.

A CIP catalogue record for this book is available from the British Library.

All rights reserved. No part of this book may be reproduced or transmitted in any form or by any means, electronic or mechanical including photocopying, recording or by any information storage and retrieval system, without permission from the Publisher in writing.

Typeset by Mac Style
Printed in the UK by CPI Group (UK) Ltd, Croydon, CR0 4YY.

Pen & Sword Books Limited incorporates the imprints of Atlas, Archaeology, Aviation, Discovery, Family History, Fiction, History, Maritime, Military, Military Classics, Politics, Select, Transport, True Crime, Air World, Frontline Publishing, Leo Cooper, Remember When, Seaforth Publishing, The Praetorian Press, Wharncliffe Local History, Wharncliffe Transport, Wharncliffe True Crime and White Owl.

For a complete list of Pen & Sword titles please contact

PEN & SWORD BOOKS LIMITED
47 Church Street, Barnsley, South Yorkshire, S70 2AS, England
E-mail: enquiries@pen-and-sword.co.uk
Website: www.pen-and-sword.co.uk

Or

PEN AND SWORD BOOKS
1950 Lawrence Rd, Havertown, PA 19083, USA
E-mail: Uspen-and-sword@casematepublishers.com
Website: www.penandswordbooks.com

*Some people seek riches; others are in pursuit of power.
I chase history.*

                                                                           Anon

Contents

Acknowledgements		viii
Prologue		x
Chapter 1	The Augsburg Raid	1
Chapter 2	'The Unwanted Hat Trick'	24
Chapter 3	'Sheff'	42
Chapter 4	Lancaster Love Story	60
Chapter 5	Rustle of Spring	92
Chapter 6	The Valley Of Death	104
Chapter 7	"For Those Who Wait"	127
Chapter 8	Orphan Airman	153
Chapter 9	Under the Influence of Mars	170
Chapter 10	Kaiser's Crew	190
Chapter 11	Attacks on The *Tirpitz* 1944	211
Chapter 12	*Aries*	228
Chapter 13	Epilogue: Two Sides to Every Story	242
Index		249

Acknowledgements

I am indebted to Len Traynor for his wonderful prologue and to Geoff Reynolds, Secretary of the Mildenhall Register and author of *Orphan Airman* for kind permission to adapt and use his unpublished story and to Dick Breedijk, author of *Ted Robbins Bommenwerper Piloot* (for which he literally earned a medal from the Dutch Government in honour of his researches!) I would also like to thank Mike Wooldridge for his generous help in the preparation of the story about 'Kaiser's Crew' and as well, Squadron Leader Clive Rowley MBE RAF (Retd) and also the late Squadron Leader Francis 'Frank' Ridley Leatherdale DFC and his family for kind permission to quote passages from his unpublished memoir. Equally, Squadron Leader Ed Bulpet RAF (Retd) for his help and direction concerning *Rustle of Spring;* Elizabeth Morris for kind permission permitting me to quote her father's stories from *Pathfinders* by Wing Commander Bill Anderson OBE DFC AFC; John Hoyte; and the late Ronald Homes DFC AGAVA. I am most grateful to Jerry and Rosemary Nice for kindly providing me with several accounts and photos relating to the events concerning the late Squadron Leader Backwell-Smith (Rosemary's father) and Norman Sirman.

Several invaluable reference works have been very welcome and they include RAFCommands.com; Henk Welting's Database, W. R. Chorley's *RAF Bomber Command Losses of the Second World War Vol.9 Roll of Honour 1939–1947; The Bomber Command War Diaries: An operational Reference Book, 1939–1945* by Martin Middlebrook and Chris Everitt; the International Bomber Command Centre (IBCC) records; *Nacht 15–16 Marz 1944* by Fred Trendle Buch and *Luftwaffe Night Fighter Combat Claims 1939–1945* by John Foreman, Johannes Mathews and Simon Parry (Red Kite, 2004).

Last but certainly not least I am grateful to the late Flight Lieutenant 'Steve' Stevens DFC AE* (The Air Efficiency Award given for ten years' meritorious service to airmen) and his dear wife, Maureen and equally, to Steve's niece, Rita Gulliver and her husband Peter for kindly allowing me to include the transcript they made of Steve's 1943 wire recording which was used in their delightful booklet called *Steve: A True Story of WWII* which is the result of interviews with Steve and Maureen in 2011. The wire recording had been painstakingly transferred first to reel-to-reel tape, then to cassette and finally to a CD. I shall never forget my visits to meet this delightful couple in their home in my native Norwich. Maureen, who was then 94 years of age, introduced herself and her husband who at 92 she described as her 'toy boy!'

We will never see their like again.

Martin W. Bowman, 2021

Prologue

My Uncle Jack
by Leonard Traynor

Ever since I can remember I have been a voracious reader as a child I read all the classics such as *The Three Musketeers*, *Treasure Island*, *Oliver Twist*, and the rest. Over the years I have read over sixty books on Lancasters and Bomber Command and have read as many books on films and film stars, but my special interest has been modern military history. At the age of eight I learned an Irish forbear of mine served in the Union Army during the American Civil War which fired my passion for that conflict. I have a library of two and a half thousand books on the war. So after all those years I like to think I know a good author or two. Some are dull and boring.

Others like Martin Bowman fire the imagination and transport the reader to the events as they are written. The way he writes is like being in the aircraft and smelling the flak. After finishing his books I feel I am entitled to receive the DFC. I hold him and the late Charles Bruce Catton in the same league of gripping and exciting writers. Catton was a narrative American historian and journalist, known best for his books concerning the American Civil War, featuring interesting characters and historical vignettes, in addition to the basic facts, dates, and analyses.

My interest in Bomber Command was motivated by my Uncle Jack – Pilot Officer Arthur George Jackson Chadwick-Bates DFC – my mother's younger brother – who served as a rear gunner on 460 Squadron RAAF. My memories of my uncle are still fresh, like it was yesterday even though I last saw him in 1942 when he visited us, on his last leave before going overseas to become an aircrew member in Bomber Command operating against occupied Europe.

Uncle Jack was born of English parents in Wellington, New Zealand, on 16 August 1910. His parents owned a copra plantation in New

Hebrides, a French colony in the South Pacific and 'Chubby', as he was fondly known to the family, grew up in the idyllic surroundings of an island paradise and of course became fluent in French.

In his late teens the family resettled in Australia, spending some time in Victoria, before moving into a house in the Sydney suburb of Lane Cove. Being a natural sportsman, he excelled in rugby, cricket and golf, his favourite sport, which he played on a regular basis with his father at the local golf course.

Even though it was in the midst of the Depression, he managed to qualify as an accountant and secured employment with A. E. Goodwin, an engineering company at Lidcombe, a Sydney suburb. He remained with them until his enlistment into the Royal Australian Air Force on 19 July 1941, to be trained as a wireless operator/air gunner.

At the time of his resignation from his employer, they presented him with a beautiful mantel clock, a token of their appreciation for loyal service to the company, and his patriotic gesture for enlisting in the service of his country. That clock now sits on my dining room mantelpiece, and its mellow chimes are sounding through the house as I write.

Like thousands of others, uncle did his basic training at Bradfield Park located in the Sydney suburb of Lindfield, before relocating to Ballarat in Victoria to commence his training as an aircrew wireless operator. I don't know if 'Uncle Jack' had ideas of being a pilot before he enlisted, or was aware that being below the minimum height of 5 feet 7 inches, would exclude him from that position.

His surviving letters to my mother and her family, which I have, commence from this time, and he makes it clear the course was far from easy, with many not reaching the high standards required. After his successful completion of this course, he moved to Evans Head on the NSW north coast to undergo his training as an air gunner. When this was completed, he was posted overseas, leaving for England in October of 1942.

Prior to his departure, he came to visit us at our house in Barker Street, Kingsford, just across the road from the Kensington Race Course, which during the war years was not operational, but was occupied by the Australian army, which was a great attraction to my elder brother and me. We would slip through a gap in the fence and spend time with the soldiers. Many had just returned from the Middle East and enjoyed the company of us boys, who no doubt reminded them of their own children. My time with these Aussie diggers gave me a love of the army which has never left me, culminating in me serving as a National Serviceman in the early 1950s and later seven years with the Citizen's Military Forces, which is now known as the Army Reserve.

On Uncle Jack's last visit with us I can still remember the game of cricket he played with my brother and me in the backyard and dominoes on the dining room table that night after dinner. Later that night we walked down to the tram stop to say goodbye; he standing there looking resplendent in his uniform, with the overcoat over his arm and my mother with tears streaming down her face. These tears turned to heart rending sobs, as he boarded the tram and gave a final wave. My mother's heavy sobs continued on the walk home. She no doubt realised what I didn't; that maybe she was seeing her only brother for the last time. Her sorrow was contagious. I clasped her hand tightly and I cried along with her.

On the trip over and his time in England, Uncle Jack wrote on a regular basis whenever he had the time, I was thrilled that he even made the time to write to me, which excited me beyond measure and I of course wrote back keeping him informed of my progress at school and my sporting activities. I was so proud of my Uncle Jack.

After more training in England, in June 1943 Uncle Jack was eventually assigned to 460 Squadron RAAF. The family of course didn't know this, as our letters were all directed to Kodak House in London.

Unknown to us, uncle, and thousands of other aircrew faced extreme danger on a daily basis, with death as a constant companion, night after night, they clambered into their bombers, which were loaded with petrol, and high explosives, and participated in bombing raids over occupied Europe. Facing the hazards of bad weather, German night fighters, anti-aircraft, and the added threat of mid air collisions in the crowded skies over enemy targets.

Uncle Jack was in the midst of all this and saw at first hand the dangers and the appalling casualty rates resulting from these raids. Yet, in his letters home there was never any indication of the situation he endured. Wartime censorship prevented him from saying anything about it, although occasionally he does mention the name of a friend who has gone missing, or been killed on operations.

Little did we know that he had done seven trips to Berlin, two to Hamburg, and two all the way to Italy, plus of course a number of other German cities like Düsseldorf, Stuttgart and Frankfurt etc.

His letters were always positive, often writing what he would do when he came home, telling us about the things he saw in England during those times he was on leave, visiting relatives, watching sport, the weather and the occasional romance, etc. There is not one depressing note in any that he wrote to us, but what his private thoughts were at the time, we have no way of knowing.

In January 1944 I remember the excitement in the family when we received the news that he had been commissioned by being promoted to pilot officer and his letters telling us of going to London to order his new uniforms.

Rear gunners on operations had the highest casualty rate of any aircrew and their survival was measured in weeks. Uncle Jack was beating the odds; he had survived eight months. He, like all other aircrew, counted each successful operation, as it meant one step closer to completing their tour of thirty operations. Then they would be free from the stress and dangers of bombing raids and could transfer to a training situation, if they so decided, or could continue on operations.

On 27 March 1944, he wrote to us saying," I have one more trip to do and I am finished with ops". That final trip was to be the disastrous raid on Nuremburg, 30/31 March 1944, which from most accounts, was a complete an utter failure. Bomber Command lost the most bombers on a single raid during the whole war. Ninety five aircraft were lost, in the slaughter in the skies over Germany. Much has been written about this raid, which still provokes controversy to this day, as hundreds of allied airmen gave up their lives, for very little effect on that city. As a long time student of Bomber Command I can say with all sincerity that Martin Bowman's book is the best I have ever read on that subject; so descriptive and so detailed it left me marvelling the amount of research involved in its preparation.

On that night, Uncle Jack's run of luck came to an end, about 1.30 am on 31 March, on their approach to the target, uncle's aircraft, Lancaster ND361 piloted by Squadron Leader Eric Arthur Gibson Utz DFC*, was attacked by an enemy night fighter causing the aircraft to crash, killing all the crew except one, who managed to parachute to safety.

Back home in Australia, we were unaware of this tragedy, we knew from newspaper reports that heavy losses had occurred on the raid, but Uncle Jack's letter had not arrived. But in early April his mother received the dreaded telegram couched in official language: "your son, 412480 Pilot Officer A. G. J. Chadwick-Bates, is missing as result of air operations" etc, etc.

This devastating news was conveyed to us by a telephone call, and it was like a hand grenade being lobbed among the family. We were all shattered, distressed as she was, my mother held out hope that he may have survived, and was a PoW. Not long afterwards his letter arrived, advising he had one more trip to go etc. This letter only compounded the stress the family was experiencing, as no other official news had been received.

Months later, a trunk arrived at his parents' house, sent by the "Committee of Adjustments", and no doubt containing my uncle's uniforms and personal effects… normal procedure for aircrew members killed or posted missing on operations. His parents were too distraught to open it; it was just packed away.

The weeks turned into months with no further news, with the family living on a knife edge. This period affected my mother greatly, as it did for the rest of the family, his parents and younger sister. My mother was prone to periods of heavy weeping; it was not a happy household, as we tried to get on with our lives, making it as normal as circumstances would permit.

Finally, the following September we received the shattering news that my dear Uncle Jack had been confirmed killed. On receipt of this news, my mother didn't cry, she screamed like a banshee, screamed and screamed: to think of it now still chills my blood.

His loss affected all us very badly, my mother never recovered from the death of her only brother and in the decades following his death, she rarely mentioned his name. Any attempt on my part to ask a question about him was always met with a outpouring of tears, which discouraged me from enquiring any further, which was in a way, was very regrettable, as it prevented me from finding out more about uncle and his life before the war.

Somehow she was not convinced he was dead. Until the day she died some forty years later, she hoped that one day he would walk through the front door. Sadly that day never came.

Not long after Uncle Jack was posted missing, we read in the newspaper that he had been awarded the Distinguished Flying Cross. The family although very proud of the award, thought it was poor compensation for the loss of their son, and brother.

I remember sometime in 1946, my grandparents attended Government House in Sydney and were presented with their son's decoration, by the then Governor-General, the Duke of Gloucester. When we visited my grandparent's house, I used to look at the DFC, in its white satin lined box, sitting on the mantelpiece in the lounge room. For some reason, it was not until 1953 that my grandfather applied for his son's campaign medals, but I don't ever remember seeing them.

It was not until my grandmother died in 1958 that the trunk sent home all those years before, was finally opened by my mother. It contained uncle's uniforms, personal effects, and the many photographs he had taken while on leave. It was never explained to me, but much of what the trunk contained was thrown out by my mother, maybe seeing it all just created fresh grief. Somehow two tunics, his overcoat, gunner's mittens, photographs, a couple of manuals, and his log book survived the purge and I treasure them.

Reading his log book and seeing the raids in which he had participated increased my admiration for him and all other members of aircrew who had faced the same dangers. I was surprised that for some reason Uncle Jack was not a part of a regular crew; he was 'an odd bod' – a person who fills in for other gunners, who were either sick, wounded or KIA. Of the 32 operations he had flown, he flew 13 bombing raids with Squadron Leader Carl Kelaher (KIA, 4 September 1943), one with Wing Commander Robert Norman, four with Warrant Officer Wilfred Munsch, two with Flight Lieutenant Eric Greenacre and ten with Squadron Leader Eric Utz, including the fatal raid on Nuremburg. On one notable occasion, on a trip to Berlin in July 1943, Uncle Jack's pilot was none other than Group Captain Hughie Edwards VC DSO DFC the 460 Squadron CO.

As I became older, Uncle Jack's service prompted me to read about Bomber Command and over the years I have accumulated a small library about this branch of the service. From the reminiscences of former air crew, I learnt more about the terrors, and stresses of being a member of a squadron operating against occupied Europe, my admiration for these men increased.

From Canberra, I obtained copies of my uncle's service record, and information where he, and his crew that were killed in the crash are buried, the Hanover War Cemetery. Sometime in the 1970s, unbeknown to my mother, I wrote to the War Graves Commission in Germany and obtained a photograph of Uncle Jack's headstone. I never told her I had it, as I knew she would be extremely upset.

In March 1997 I travelled to Hanover and visited my uncle's grave. It was cold and wet on the day of my visit, in keeping with my sombre mood. With the help of the information I had received from Canberra, and the assistance of a cemetery worker, who spoke excellent English, there amongst a sea of white headstones, I found uncle's grave, he lies there side by side with those crew members who died with him, a very emotional moment I can assure you.

As I walked among the rows of headstones, reading the names and other inscriptions carved into the stone. I was struck by the youth of so many of these men who had made the supreme sacrifice to preserve our freedom and reflected on the terrible loss of life that had occurred in the skies above Europe during those dark days of 1939–1945.

I returned again the next day, to spend more time at my uncle's grave and recall my memories of him all those years ago and regret that so many of our present generation have no idea, of what the men in bomber command experienced, or the vital role they played in winning the war.

Over the years I had wanted to know more about the details about the loss of Lancaster ND361. Some information came from Canberra. Only recently through the good graces of Richard Munroe of the 460 Squadron Association, who put me in touch with the son-in-law of Pilot Officer Ronald James McCleery DFC the navigator who had parachuted to safety from the crashing bomber, I learned that he was hit in the leg but bailed out and was about to be lynched by villagers when a local doctor intervened. McCleery later had the leg amputated. All this came as a great surprise to me, as I was under the impression all the crew had been killed in the crash. McCleery died in 1959.

His son-in-law had visited the crash site and even spoke to local people who remember the Lancaster crashing, and the aftermath of the event. He even obtained a portion of the crashed bomber and very graciously gave me a piece, which he had very nicely mounted on a base with the names of the crew engraved on a plaque, along with some photographs of the crash site; items to treasure.

The mementos I have of my uncle, plus my memories, are all that remain of this brave young man, who served his country in her hour of need, and like thousands of others made the supreme sacrifice.

I am very proud to be his nephew.

Wentworthville
New South Wales
Australia
March 2020.

Chapter 1

The Augsburg Raid

Officers' Mess Royal Air Force Waddington,
Lincolnshire.
Tel. Waddington 464
Tuesday, April 15th 1942
My darling mother,

I knew from the start that this was bound to happen in the end, and I have always thought that my only regret would be not saying thank you and good bye. It seems strange writing so but I feel. I must.

I will not begin to thank you for everything because words cannot express it, and anyway it would take far too long.

Like Dad I am not afraid to die but just don't want to. But 'God's will be done' so instead of coming home to you I go and meet Dad. I have heard how brave you were when Dad left us so I have no fear now. It is rather strange I should mention those words about God's will as I remember so well when I last said them. I was just before I passed out when in the sea that time.

I should have loved to see you all once again Tricia, Guy and little Anthony, but that's how it must be, I suppose. You are a wonderful 4 and I hope you stick together and see Hitler beaten. But for you people there would be nothing in life to live for.

Dear 'Fatty', I wonder if he really remembers me or if he has just heard you talk of me so often. Actually I think he does remember me – perhaps in the swimming bath. God, what a home-life I have had. Everything a man could wish for and I don't think I appreciated to the full.

You remember, Mum how I used to say, 'When I find a woman like you I would marry her tomorrow'. I now realize that if I had kept to that and if I had lived, I should never have been married...

Letter written to his mother by 21-year-old Rhodesian second pilot, Hurworth Anthony Paul ('Buster') Peall on 44 (Rhodesian') Squadron the day before he set out on the Augsburg raid. The letter was only to be posted in the event of his death.

In 1941 44 Squadron at Waddington a little over four miles south of Lincoln had been renamed the 'Rhodesia' Squadron in honour of that colony's contribution to Britain's war effort. It was also to recognise that up to 25% of the ground and air crew were from Southern Rhodesia. In the Great War Rhodesia had contributed almost 6,000 men to the Allied war effort and could ill afford the seven hundred dead. Among those living in Rhodesia in 1913, was a man by the name of Arthur Harris who born on 13 April 1892, at Cheltenham, Gloucestershire but had emigrated to Southern Rhodesia in 1910, aged seventeen. He joined the 1st Rhodesia Regiment but he returned to England in 1915 to join the Royal Flying Corps, with which he remained until the formation of the Royal Air Force in 1918. No.44 Squadron had been formed at Hainault in Essex on 24 July 1917 and by the end of the war it was commanded by Harris. He remained in the Air Force throughout the 1920s and 1930s. When in February 1942 he assumed command of RAF Bomber Command, among the squadrons he inherited was 44 Squadron, which became the first to convert completely to Lancasters, flying their first operational sorties in this aircraft on 3 March 1942 and 97 (Straits Settlements) Squadron, which became the second Lancaster squadron after beginning conversion in January 1942. Four aircraft on 44 Squadron made the first Lancaster operation of the war with a mine-laying sortie in Heligoland Bight on 3 March 1942. The first night-bombing operation was on 10/11 March when two of the Squadron's aircraft took part in the raid by 126 aircraft on Essen.

On 11 April 1942 Lancaster crews on 44 Squadron and 97 Squadron at Woodhall Spa sixteen miles south east of Lincoln probably thought that ops would be 'on' again that night but, without being told the reason, they were ordered to prepare for an epic raid and it involved making long-distance flights in formation in daylight in order to obtain endurance data on the Lancasters. Two days later ground crew at Waddington were instructed to prepare eight machines to move to an advance base. Similar instructions were given to 97 Squadron at Woodhall Spa. Crews on both the squadrons knew that the real reason was that they were training for a special operation code named 'Margin') and speculation as to the target was rife. On 15 April, the Lancasters, flying in groups of three in 'vic' formation at extremely low level made their way south to Selsey Bill. There they turned and headed north to Lanark, across country to

Falkirk and on to a point just outside the town of Inverness where they feigned an attack. They then returned to base following the route they had taken on the way north. A spate of complaints from all over the country was received by the Air Ministry castigating the pilots for the dangerous prank of flying large four-engined aircraft at treetop height over the countryside. A tactful letter was sent to all police stations but the aircrews of both squadrons already realized that they were in training for a special operation and bets were placed on the possible target.

'Despite frequent groundings', recalls 23-year-old Flight Lieutenant David Jackson Penman DFC, otherwise known as 'Jock', a pilot on 97 Squadron, 'training continued and early in April rumours of some special task for the Lancasters were confirmed when eight crews were selected to practice low level formation flying and bombing. The final practice was a cross-country at 250 ft for two sections of three led by 23-year-old Squadron Leader John Seymour Sherwood DFC* with myself leading the second section.' The son of a First World War Army officer, Penman was born in Edinburgh and educated at the Royal High School. On leaving school in 1937, he was granted a short service commission in the RAF to train as a pilot and he had joined 44 Squadron at Waddington in October 1938.

'We took off from Woodhall Spa and were to rendezvous with 44 Squadron near Grantham but because of un-serviceability they did not take off. We flew down to Selsey Bill and then turned round and headed for Inverness. Due to compass errors the lead section got off track and they were heading into an area of masts and balloons. With no communication allowed I eventually parted company with the lead section and we did not see them again until we were bombing the target in the Wash at Wainfleet. Our low-level flight up valleys to Edinburgh was exciting, but over the higher ground in the North we climbed to a reasonable altitude over cloud, descending in the clear at Inverness for a low-level run. Once beyond Edinburgh, on the way back, we descended again to low level and, full of confidence, really got down to hedge hopping."

Flying Officer Ernie Deverill on the left in 'Y-Yorker' and 22-year-old Warrant Officer Thomas James Mycock DFC on the right in 'P-Peter' maintained very tight formation. The 26-year-old Ernest Alfred 'The Devil' Deverill DFM was a Halton 'Brat' (ex-apprentice) who undertook pilot training in 1938. He had flown over a hundred operational sorties, mostly in Coastal Command and had been awarded the DFM for assisting the wounded pilot of their damaged Hudson to return with a dead gunner after an attack by three Me 109s.

'My only regret was the stampeding cattle when we could not avoid flying over them. Greater satisfaction came as we roared across familiar airfields a few feet from the hangar roofs and Waddington got the full blast of our slip stream as we rubbed in our success whilst they were stuck on the ground. A perfect formation bombing run with Sherwood's section running in behind completed a very successful day.'

A few days later 'Jock' Penman went to HQ 5 Group in Grantham with the Station Commander from Coningsby and Squadron Leader Sherwood. 'At 5 Group when the target was revealed, we were shattered and suicide was common thought' recalled Penman. The target, code-named Operation Margin' was the diesel engine manufacturing workshop at the Maschinenfabrik Augsburg-Nurnberg Aktiengesellschaft (M.A.N.) factory at Augsburg the other side of the Danube, which would involve a flight of 1,500 miles over enemy occupied territory. Half the Diesel engines for ocean-going U-boats were being produced in the M.A.N. factory. To reduce the supply of diesel engines would throw the whole submarine-building programme out of gear. At a time when the U-boats in the Atlantic were becoming a most serious menace it would be hard to imagine any more important single target than this assembly shed at Augsburg. It would be worth serious losses to hit this target.

"However, the briefing was thorough with an excellent scale model of the target area and emphasis on low level to avoid detection, massive diversionary raids and little ack-ack or opposition at Augsburg. This briefing was only a day or two before the 17th and no one else was to be informed until the briefing on the day of the raid when take-off was to be 1515 hours. On Friday the 17th briefing was immediately after lunch with crew kitted ready to go. The scale model of the target was on display and the gasps as crews entered the room and saw the target were noticeable."

At Waddington Wing Commander Learoyd vc the 44 Squadron CO, began his address to the crews with 'Bomber Command have come up with a real beauty this time' and added 'I shan't be coming with you. I've got my VC already. I've no desire to get another.' Roderick Alastair Brook

'Babe' Learoyd had been born at Folkestone on 5 February 1913. He was awarded Bomber Command's first Victoria Cross of the war for his determined attack on the Dortmund-Ems Canal on 12/13 August 1940 in a 49 Squadron Handley Page twin-engined Hampden.

At Woodhall Spa when the target was announced at the briefing the Intelligence Officer did not receive the response he must have been expecting. 'When the curtain drew back there was a roar of laughter instead of a gasp of horror. No one believed that the air force would be so stupid as to send twelve of its newest four-engined bombers all that distance inside Germany in daylight' recalled Flying Officer Eric E. 'Rod' Rodley, who was known for his whimsical sense of humour. 'We sat back and waited calmly for someone to say "Now the real target is this". Unfortunately it was the real target, a factory near Munich that was a major manufacturer of diesel engines for submarines."

Before the laughter had died down, Wing Commander John David Drought 'Joe' Collier DFC the 97 Squadron CO entered and walked quietly forward to the front of the briefing room and mounted the dais. The crews came to order at once, listening intently. Collier was born near Plymouth on 10 November 1916, the son of a businessman. After education at St. Petroc's School in Bude, Cornwall and Tettenhall College, Staffordshire he trained as a land agent under Lord Leigh at Leamington Spa. Subsequently he was employed by John Bishop of Northam, Devon, whose daughter Elizabeth he married in 1939. In 1936 Collier received a short-service commission and was posted to 83 Squadron at Turnhouse in Scotland where he flew Hawker Hind biplane light bombers. In 1938 he moved with the squadron to Scampton, Lincolnshire, where it was re-equipped in No 5 Group with Hampdens, known on account of their elongated appearance as 'panhandles'. In 1940 Collier was a flight lieutenant on 83 Squadron, flying Hampdens, often on one named 'Bet' in honour of his wife. A particularly hazardous operation took place on the night of 12/13 August when Collier led three Hampdens in a diversionary attack on Münster, while other members of 83 Squadron, under Roderick 'Babe' Learoyd, attacked an aqueduct. In the same period Collier led a low-level attack on an oil refinery at Bordeaux, for which he was awarded the first of his two DFCs. Despite intense fire from local defences, Collier managed to set the tanks ablaze, so that the enemy gun positions were also enveloped in flames.

'Well gentlemen', smiled Wing Commander Collier, 'now you know what the target is'.

'Rod' Rodley was philosophical about it all when the target was known. "At that time it was touch and go in the North Atlantic between Britain having enough to eat and not having enough to eat. The crews were determined that these diesels should not go forth in submarines."

However, when all said and done, self-preservation concentrated minds more fully and Rodley and his crew of 'F-Freddie' clung to the hope that they were only the reserve. Rodley had been 'one of the lucky ones', joining the RAFVR in 1937 and he already had instructing experience when war broke out. Consequently, he graced the first 'War Instructors Course' at CFS in October, 1939, after which he spent an exhausting and unrewarding year and a half in Training Command. Hence, by June 1941 he had over 1,000 hours in his logbook but he was 'quite unprepared for any wider aspects of the art.' A strange set of circumstances rescued him from the thraldom of instructing when 97 Squadron was re-formed and equipped with the Avro Manchester when all pilots had to have at least 1,000 hours under their belts.

But Rodley and the crew of 'F-Freddie' would not be the reserve. "The route took us low, at about 100 ft, down to the south coast, across the Channel. We were to join 44 Squadron at the south coast, six aircraft from each squadron, and we were to go as a formation of twelve the rest of the way."

Air Marshal Harris wanted the plant raided by a small force of Lancasters flying at low level (500 ft) and in daylight despite some opposition from the Ministry of Economic Warfare, who wanted the ball-bearing plant at Schweinfurt attacked instead. Crews were ordered to take their steel helmets. Sixteen Lancaster crews, eight each from 44 (Rhodesia) and 97 Squadrons (including four first and second reserve) were specially selected. Squadron Leader John Dering Nettleton, still on his first tour in 44 Squadron, having spent much of the war to that time as an instructor, was chosen to lead the operation. The 24-year-old South African had been born at Nongoma in Natal and educated in Cape Town. The grandson of an admiral, he had spent eighteen months at sea in the merchant service before coming to England to join the RAF in 1938 to train as a pilot. Dark haired but fair-skinned, tall and reticent, grandson of an admiral, he was an inspired choice as leader.

Pilot Officer Patrick Arthur Dorehill, Nettleton's 20-year-old second pilot from Fort Victoria in Southern Rhodesia had aviation in his blood ever since he had watched the weekly Imperial Airways flying boats and South African Junkers passenger planes flying over his native land. He was studying mining engineering at Rand University when war broke out and briefly worked in coal mines before joining the RAF in July 1940 and training as a pilot. 'There was certainly some surprise on entering the briefing room to see the pink tape leading all the way into the heart of Germany' he wrote. 'I can't say I felt anxious. I had an extraordinary faith in the power of the Lancaster to defend itself.' Dorehill was confident in their new aircraft. "I had done fifteen operations on Hampdens and before that I had always been nervous, but flying low level, with all the armament the Lancaster carried, I thought it would be a fairly easy affair. I thought six Lancasters, with all the armament we had, we would be a match for any fighter. And then flying at low level seemed to me to be the perfect way to outwit the enemy. I thought the only danger might be over the target and, even there, believed we would be in and away before there was much response.'

One of the pilots on 97 Squadron was Norfolk-man Flying Officer Brian Roger Wakefield 'Darkie' Hallows, who after flying three days of long formation cross-countries was set to fly 'B-Baker'. Hallows was eleven years at Gresham's School, Holt, where his father was proprietor of the old Norfolk market town's steam laundry. He later passed through Sandhurst and took a commission in the King's Liverpool Regiment, which he later relinquished to take up flying. In June 1938 he joined the RAFVR when he felt that a war was inevitable, so it was no surprise when all Reservists were called to report to their units around the end of August 1939. Then followed a wait until October, when he begun a 'Wings' course at No.9 FTS Hullavington. They were an unruly lot and finally the station commander got them all together and said that he was going to award them their wings and commission all sergeant pilots to acting pilot officers. He added that they were not fit to be senior NCO pilots – how right he was! Then it was off to Central Flying School, to become a RAF flying instructor, which had been Hallows' profession prior to August 1939. At CFS he won a trophy for 'Best all-round Cadet' In June 1941 he had seen a 'rather odd' notice asking for experienced twin-engined pilots to join Bomber Command, to fly Manchester bombers.

Little did any of them know what a killer the Manchester was. Anyway, it was off to Finningley, with not a Manchester to be seen – they were all grounded with engine trouble. Eventually, in September 1941, Hallows got to 97 Squadron at Coningsby, where he actually flew a Manchester.

It was Hallows' reputation for the use of R/T language 'which caused some watch tower WAAFs to giggle, some to blush' and not for his jet black hair and full moustache that earned him his nickname 'Darkie'. When he had got lost and invoked the R/T get-you-home service of those early days: 'Darkie, Darkie', receiving no response, he had tried again, but still no reply. Once more he had transmitted to the void: 'Darkie, Darkie – where are you, you little black bastard?'

'Plenty was said about how important it [Augsburg] was and all that stuff' recalled Hallows. 'So we were obviously not intended to come back in any strength. Fighter Command had been on the job for several days, hounding the German fighters and when we were on the job we saw no fighters at all, all the way…'

Since the Lancasters could not be escorted to the target, to avoid radar detection they would fly at low altitude all the way to Augsburg. Fighter Command would put on a large diversionary 'Circus' operation just before the Lancasters took off to draw the enemy fighters into combat so that the passage of the Lancasters would coincide with their refuelling and re-arming. A 'Ramrod' style attack would be carried out by a force of thirty Boston medium bombers on 88 and 107 Squadrons in 2 Group accompanied by a heavy fighter escort. One group of bombers would attack the port facilities at Cherbourg while a second group attacked the shipyard at Rouen to the east. Meanwhile, large 'Rodeo' operations would be conducted to the north in the Pas de Calais. These diversionary attacks were intended to divert the German fighters away from the entry point for the Lancaster force, allowing them to get past the coast defences and into northern France. Unfortunately it had the opposite effect and the incursion put the Luftwaffe on the alert.

For take-off the duty aerodrome control pilot, Sergeant Knight, had to be aware of the timing of the raid and in the cookhouse for duty ration detail, Aircraftman Shaw, preparing sandwiches and flasks, had to know how many crew would be involved and roughly how long the operation would last. It would be dark when the crews returned and so the officer in charge of night-flying, the duty flare path assistants and floodlight operator, together with the duty electrician, Aircraftman Saunders, had to be aware of possible timings and numbers of aircraft involved in case of snags. An operation of this importance involved the whole squadron.

The Lancasters on 44 Squadron were to form the first two sections of the flight. At 1512 hours Nettleton lifted 'B-Baker' off the Waddington runway to be followed by 'A-Apple' his No.2 flown by 22-year-old Flying Officer John 'Ginger' Garwell DFM to starboard and 'H-How' his No. 3 piloted by 29-year-old Sergeant George Thomas 'Dusty' Rhodes from Leeds, to his left. The second 'vic', close behind, was led by 24-year-old Flight Lieutenant Reginald Robert 'Nicky' Sandford DFC in 'P-Peter', with 'V-Victor' skippered by 22-year-old Warrant Officer John Frank 'Joe' Beckett DFM RAFVR, who was from Oxford, and 'T-Tommy' piloted by Warrant Officer Herbert V. Crum DFM, a wily bird, old in years and experience by comparison with most of the others.

Once the formation was airborne and beginning to close up, the last Lancaster to depart – a reserve bomber – circled and returned to base. The six aircraft flew low across Lincolnshire heading south. The Lancasters departed Selsey Bill on schedule. The April sky was clear, the day warm.

Ten miles to the east, at Woodhall Spa, the six Lancasters of 97 Squadron, led by Squadron Leader Sherwood, were also lifting off and forming up. The other five aircraft in the group were U-Uncle' piloted by 'Jock' Penman and 'B-Baker' being flown by 'Darkie' Hallows and 'F-Freddie' piloted by 'Rod' Rodley, 'Y-Yorker' skippered by Ernie Deverill and 'P-Peter' captained by Warrant Officer Tom Mycock.

'Take-off was at 1515 hours' recalled 'Jock' Penman "with the two reserve aircraft taking off and dropping out when the two 'vics' of three set course. [All sixteen taxied out so the second reserve cut their engines and watched the others leave and over Selsey Bill the first reserves swung round and returned to Lincolnshire, leaving twelve Lancasters [Call sign 'Lifebuoy'] flying low over the water. We were to meet 44 Squadron [Call sign 'Maypole'] near Grantham and then on to Selsey Bill, across the Channel, then down south of Paris before turning left and heading for Lake Constance. Take-off was on time, singly, with full fuel [2,134 gallons] and four 1,000lb RDX bombs with eleven-second delay fuses to be dropped from 250 ft. Weather forecast was perfect with clear skies and good visibility all the way. I took off and soon had Ernie Deverill on my left and Tom Mycock on my right. We joined the lead section of

Squadron Leader Sherwood, Flying Officer Hallows and Flying Officer Rodley. As they had watched the other six aircraft starting up some of Rodley's crew had begun to have mixed feelings and almost wished now that they had got this far that they were going.

"Then 'A-Apple' had a mag drop on No.1 engine and Rodley's crew had fallen silent. Rodley thought of his wife in Woodhall Spa. She would probably guess what had happened. He took off with the others and moved into the No.3 position, tucked in to port of Sherwood. Once again there was no sign of 44 Squadron near Grantham and we were never to meet. [As the bombers hugged the waves Nettleton's sections began to draw ahead of Sherwood's formation, flying slightly off the intended flight path. Sherwood made no attempt to catch up; the briefing had allowed for separate attacks if circumstances decreed such and Sherwood was highly conscious of the need to conserve fuel on such an extended sortie]. 'We maintained 250 ft to Selsey Bill and then got down as close to the sea as possible for the Channel crossing. As we approached the French coast my rear gunner informed me his turret was U/S and I told him it was too late to do anything about it and he would just have to do what he could with it. Crossing the French coast was an anti-climax as not a shot was fired and we flew on at tree top height to Lake Constance. We saw the odd aircraft in the distance but otherwise it was a very pleasant trip.'

'Penman's Australian navigator, Pilot Officer Edward Lister 'Ding' Ifould saw frightened bullocks scampering across the furrows with their ploughs bumping along behind them. He also saw French workers wave to them but usually only when they were working in secluded parts where their greeting would be unobserved. One greeting came from two or three workers deep in a quarry, down which Ifould had a momentary glimpse as their Lancaster flew over almost at tree-top height; and again, in a wood, some charcoal burners stopped to give them a secret wave. Ifould, a keen sportsman, five feet 8 inches tall, with brown eyes and hair, strong features and a dark complexion, was born on 6 April 1909 at St. Peters, Adelaide, and the eldest of three sons of South Australian-born parents. In 1926 he joined the Colonial Sugar Refining Co. Ltd and he worked as a production chemist in the company's factories in North Queensland and New South Wales. 'Ding' enlisted in the RAAF on 22 July 1940 under the Empire Air Training Scheme. Commissioned on 14 March

1941 as an air observer, he arrived in England in July and undertook operational training before being posted in December to 97 Squadron.

The Lancs passed over no fewer than twenty-seven German airfields almost at ground level and saw nothing except a few parked aircraft. This hedgehopping was a severe test of skill and endurance. 'Jock' Penman had blisters on both hands when the job was done.

44 Squadron had not been as fortunate as 97 Squadron. Nettleton had taken his two sections flying in 'vics' of three down to just 50 ft over the waves of the Channel as the French coast came into view. From there the two formations had flown east of Caen at 1645 hours, Nettleton leading his crews down to twenty to thirty feet, so low that they had to ease the aircraft up to clear trees in their path. The bombers were flying over wooded, hilly country near Breteuil when they encountered anti-aircraft fire. Lines of tracers reached up to meet the speeding Lancasters and black shell bursts dotted the sky around them. Shrapnel from the flak bursts ripped into two of the aircraft, but they held their course. Warrant Officer Beckett's Lancaster was the most seriously damaged, its rear gun turret put out of action.

As Nettleton's six aircraft, now well ahead of the 97 Squadron formation skirted the boundary of the Beaumont le Roger airfield, just 60 miles from the coast, they ran into trouble. As the bombers appeared a group of Messerschmitt 109 and Focke-Wulf 190 fighters of 2 Gruppe/Jagdgeschwader 2 "Richthofen," one of the most experienced fighter units in the Luftwaffe, were returning to base after an engagement in the Cherbourg area with some of the diversionary RAF aircraft. For a moment the Lancaster crews thought they had not been spotted, but then several German fighters were seen to turn in the direction of Nettleton's two sections and a running fight lasted an hour. The Lancasters tightened formation, flying wingtip to wingtip to give mutual protection with their guns as they skimmed low over villages and rising and falling countryside. The Bf 109s were forced to attack from above.

'It was only sheer bad luck' wrote Patrick Dorehill 'that we flew past an enemy airfield [Beaumont le Roger] to which their fighters were returning from the diversionary raids our fighters and Boston bombers had laid on to the North. 'Up they came and I shall never forget those terrible moments. I do not think there were as many fighters as our gunners reported; it was just that each made several attacks which made it seem like more. Being on the jump seat I stood up and saw quite a bit of the action. Maybe there were a dozen. At any rate I looked back through the astrodome to see 'Nick' Sandford's plane in flames. A little fellow with a pleasing personality, who was keen on music and who bought all the records for the Officers' Mess, he always wore his pyjamas under his flying suit for luck. He thought it would bring him good luck.'

Sandford had faced the milling enemy fighters alone and forced P-Peter' down even lower in a desperate attempt to shake off his pursuers. Suddenly ahead of him he saw a line of telephone wires. He held the nose down and flew underneath them but the FW 190s followed firing all the time. This time Sandford was all out of luck and had all four engines set on fire by Feldwebel Bosseckert. 'P-Peter' crashed into the ground at Ormes and exploded in a giant fireball with the loss of all the crew. Among the dead was 'Buster' Peall, the second pilot who had written a last letter to his mother the day before. Another was 21-year-old fellow Rhodesian, Sergeant Peter Johannes Venter, one of the two wireless operator-gunners.

Nettleton saw two or three fighters about 1,000 ft above. 'The next thing I knew, there were German fighters all round us. The first casualty I saw was Sergeant Rhodes' aircraft. Smoke poured from his cockpit and his port wing caught fire.' 'H-How' went down to the guns of Spanish Civil War veteran Major Walter 'Gulle' Oesau, Kommodore, Gruppe II./Jagdgeschwader 2, a 100-victory ace officially forbidden to fly more operations, but who had jumped into a fighter and taken off on first sight of the Lancasters, followed by his wing man Oberfeldwebel Edelmann. None of Rhodes' crew stood a chance at such a low altitude in an aircraft travelling at 200 mph. With every gun on the Lancaster jammed, they were unable to fight back when attacked by the two 109s. Both port engines erupted in flames, the fire spreading instantly to the starboard engines. 'H-How' reared up, stalled and plunged straight down in flames passing between Nettleton and 'A-Apple' flown by 'Ginger' Garwell in a vertical dive and missing both by a fraction.

'He came straight for me out of control' recalled Nettleton 'and I thought we were going to collide. We missed by a matter of feet and he crashed beneath me.'

Patrick Dorehill could see the faces of the doomed crew in the cockpit. 'It was quite gruesome' he wrote later. 'H-How' crashed vertically at Le Vieil-Évreux with no survivors.

Two others went down almost at once and Nettleton saw a fourth on fire. 'At the time I was too much occupied to feel very much. I remember a bullet chipped a piece of perspex, which hit my second pilot in the back of the neck. I could hear him say 'What the hell.' I laughed at that.'

44 Squadron's flight path had taken them close to the Luftwaffe airfield at Beaumont-le-Roger and the second 'vic' was quickly shot out of the sky. 'T-Tommy', Crum's Lancaster, flew along the perimeter track as three German fighters were landing. The Lancaster crews hoped they had not been seen, but when the 109s ended their descent, wheels were retracted and they climbed away from the airfield. It was obvious they had been spotted. In the rear 'vic' Warrant Officer 'Joe' Beckett and crew on 'V-Victor' were the first to be attacked. Crews had been told at the briefing to observe strict radio silence except in emergency and Beckett had no hesitation. 'One-oh-nines!' he exclaimed over the R/T, 'at eleven o'clock high!' He had attracted most of the enemy fighters. There were about thirty of them at around 5,000 ft and the order to the Lancasters was to 'close up and keep tight'. Beckett's crew never stood a chance. His rear turret was put out of action and hit by a hail of cannon shells, 'V-Victor' became a mass of flames, dived into a clump of trees and disintegrated when it hit the ground at Le Tilleul-Lambert. Hauptmann Karl-Heinz 'Heine' Greisert of II./JG 2 was later credited with shooting the Lancaster down.

Sergeant Bertram Arthur Dowty the 19-year-old front gunner on 'T-Tommy' had seen Beckett's aircraft crash in a ball of flame as tracers flashed by and was surprised to see more 109s below as he tried to pick them off. Then 'T-Tommy's fuselage was ripped apart by cannon fire. Bullets struck Crum's cockpit's canopy, showering him and his Southern Rhodesian flight engineer, Sergeant Albert Denis Charles Dedman with shards of Perspex. Dedman looked across at his pilot and saw blood streaming down his face, but when he reached over Crum just grinned at him and waved him away. The Lancaster's guns blazed back and the two men got a glimpse of the oil streaked belly of an aircraft as the 109 flashed by and was gone. Bert Dowty recalled sea spray from the aircraft hitting the front turret as Crum lifted the nose of their Lancaster to clear the French cliffs at Le Havre. About seventy miles inland, they flew over an airfield where three Bf 109s were landing. Unteroffizier Walther Pohl of II./JG 2 flying his Bf 109, 'Black 7' singled out 'T-Tommy' and he set two engines on fire. Dowty noticed that the ground was getting uncomfortably close and he was unable to hear what was happening. 'My God!' he thought, 'we're going to crash. 'Crummy's going to fly us straight into the deck!' Dowty instinctively drew his knees up and ducked as the Lancaster flew under a high tension cable and seemed to glide, just above the stall. Bert Crum had reacted splendidly to avoid a tree and make a wheels-up belly landing in a reasonably flat wheat field at Folleville. The Lancaster tore across the field and came to a stop. The crew, badly shaken and bruised, broke all records in getting out of the wreck, convinced that it was going to explode in flames, but the fire in the wing went out. Unbeknown to Crum his crash was recorded in the Jagdgeschwader's Game Book as its 1,000th claimed victory of the war.

Dowty had avoided injury by keeping his feet fixed in the frame above the bomb sight. However, he was only too aware that they had taken off with four 1,000lb bombs and these had only 11-second delay fuses. He was to learn later that the bombs had been jettisoned before the aircraft crash-landed, but with the absence of any intercom Dowty was unaware of this. Removing the left-hand Browning machine-gun from its mounting he was feverishly trying to smash the perspex of his turret when an axe blade suddenly appeared on the left hand side. Bert Crum smashed a hole in the perspex and dragged Dowty out on to terra firma. Taking an axe from the bomber's escape kit, Crum gashed holes in the fuel tanks and tried to set fire to the escaping petrol using a Very pistol but it would not ignite. Sergeant Nicol Thomson Birkett the navigator endeavoured to set the fuel alight with incendiary canisters but this equally proved fruitless. But then Dowty remembered that he kept a 4oz 'Tom Long' tobacco tin in his turret and in the tin were ten cigarettes and a box of matches. Returning to the turret he re-covered the matches, lit two and threw them on the fuel. Nothing happened. He struck another pair of matches and this time the fuel went up, singeing his hair and eyebrows. Crum turned to Dowty and said, 'You know the drill. Destroy the kit and clear off. I want to see what's happened over there.' He had seen a blazing Lancaster about 500 yards away. Thinking it was 'V-Victor' piloted by 'Joe' Beckett his closest friend, Crum sped off to try and help but the stricken Lancaster was 'P-Peter', the last aircraft of the second 'vic' to be hit.

Crum was eventually captured. Meanwhile the remaining six members on his crew took stock. The time of the attack was evident as Bert Dowty's watch had stopped at 1712 hours. Sergeant John Miller's face was covered in blood caused by a cannon shell which had passed through his mid-upper turret grazing his scalp en route. Using the aircraft first aid pack a crewmember bandaged up his face as best he could. Sergeant A. Cobb the rear gunner had damaged his knee and was unable to walk properly. The six men then made for a wood about half a mile away and here they decided to split up. The two wounded were told to lie low in the woods while Dedman, Birkett, Flight Sergeant James Saunderson the bomb aimer and Dowty took a compass reading and headed due south. Later in their journey they took a train from Le Mans with their French guide to Tours where they alighted and boarded a bus for Châteauroux to catch a train for Limoges. Boarding the train with the tickets their guide had bought, they split up and sat in different compartments. Unbeknown to them, the secret police had boarded the train ostensibly looking for refugees making for Spain. They were taken off the train at Limoges and after a night in the police cells they were taken under escort back to Châteauroux and taken to the town hospital where to their surprise they were reunited with the two wounded members of their crew, Sergeants J. Miller and A. Cobb, who had been caught the previous night. The crew numbered six again and the following evening they were taken by train to Toulouse and later transported to Nice where they were driven along the Grande Corniche until they arrived at a fortress complete with moat and drawbridge. The fortress was to be their prison camp.

'The rest went down except Garwell on our port side' continues Patrick Dorehill. 'There was nothing for it really but to press on. A passing thought was given to turning south and then out to the Bay of Biscay but we reckoned that as we had come so far we might as well see it through. By this time I can tell you I didn't give much for our chances. On we went and I marvelled at the peaceful countryside, sheep, cattle and fields of daisies or buttercups. Along came the Alps on our right, wonderful sight, Lake Constance looking peaceful. We had climbed up a bit by then, it being pretty hilly and then down we came again getting close to the target. My recollection may be faulty but I thought we approached Augsburg from the south, following a canal or railway, factory chimneys appeared on the low horizon and then we came to the town. Large sheds were right in our path, Des Sands, the navigator and McClure the bomb aimer had done a pretty good job of map reading.[1]

'Bombs away at about a hundred feet.

'The flak zipped past and as we crossed the town to begin a left turn for home a small fire was apparent, gradually gaining strength, in Garwell's plane. Our gunners saw it make a crash landing, which seemed to go relatively well.'

Nettleton and Garwell continued to the target alone, flying low in the afternoon sun across southern Germany until the South African sighted the River Lech, which he followed to the target. Over France Nettleton had noticed people working in the fields and cows and sheep grazing and a fat woman wearing a blue blouse and a white skirt and horses bolting at the roar of his engines, with the ploughs to which they were attached bumping behind them. But once in Germany nothing was to be seen. 'The fields appeared untenanted by man or beast and there was no traffic on the roads. But when we got near the target they started to shoot at us, but the heavy flak soon stopped – I think because the gunners could not depress their guns low enough to hit us. The light flak, however, was terrific. We could see the target so well that we went straight in and dropped all our bombs in one salvo.'

Coming over the brow of a hill on to the target the two Lancasters were met with heavy fire from quick-firing guns. The bomb aimers could not miss at chimney-top height on a factory covering an area of 626 x 293 ft. Nettleton and 'Ginger' Garwell went in and dropped their bomb loads but Flight Sergeant Robert James Flux DFM his 22-year-old wireless operator yelled in Garwell's ear – 'We're on fire!'

Flux kept pointing over his shoulder and Garwell took a quick look behind him. The armour-plated door leading into the fuselage was open and he could see that the interior was a mass of flames. Garwell ordered 'Shut the door' and saw Nettleton and some of his crew staring at his burning Lancaster. Garwell stuck his fingers up in a V-sign before turning to port into wind and putting 'A-Apple' down in a field two miles west of the town. By now all five men crowded into the front cabin were coughing violently from the blinding and choking smoke and Flux opened the escape hatch over the navigator's table to try to get some air. A

---

1. Acting Flight Lieutenant Charles Curtis Granmer McClure was awarded the DFC.

sudden down draught from the hatch cleared the smoke for a fraction of a second and Garwell could see a line of tall trees straight ahead. He opened up the engines and pulled back on the stick and flew into the ground at 80 mph. The Lancaster slid on its belly for about fifty yards and the fuselage broke at the mid-turret. Garwell, his air bomber Sergeant James Watson DFM, Flight Sergeant Frank Skipworth Kirke DFM his RNZAF navigator, and the flight engineer, Sergeant Laurence Laver Dando, who was from Southern Rhodesia, scrambled from the hatch. Outside they found Flux lying dead under the starboard inner engine. He had been thrown out on impact. His quick action had probably saved their lives.[2] Air gunners' Flight Sergeant Douglas Haig McAlpine RCAF from London, Ontario and 22-year-old Sergeant Ivor Edwards, also perished.

The second formation of six Lancasters on 97 Squadron had flown a slightly different route and had avoided the fighters in France after it was thought that Nettleton was taking them north of track. The two formations planned to separate near the target and attack independently but Sherwood decided to follow his navigator's advice and the two formations had begun to separate, eventually losing sight of each other. All Sherwood's formation saw was a single German Army Co-operation aircraft, which approached them and then made off quickly.

Just inside Germany Ernie Deverill at the controls of 'Y-Yorker' noticed a man in the uniform of the SS who took in the situation at a glance and ran to a nearby post office where there was a telephone. 'Darkie' Hallows' crew in 'B-Baker' shot up a passenger train in a large station and saw an aerodrome crowded with Ju 90s. South of Paris Flying Officer 'Rod' Rodley flying 'F-Freddie' saw only the second aircraft he saw during the whole war. It was probably a courier, a Heinkel 111. It approached and recognizing them, did a 90-degree bank and turn back towards Paris. Rodley continued on flying at 100 ft. He would occasionally see some Frenchmen take a second look and wave their berets or their shovels. A bunch of German soldiers doing PT in their singlets broke hurriedly for their shelters as the Lancs roared over. Their physical exercises were

---

2. Dando and fellow Rhodesian sergeant, Donald Norman Huntley of Salisbury were awarded the DFM for "displaying …courage, fortitude and skill of the highest order". Three other Rhodesians, Pilot Officer Haworth Peall, Sergeant Pilot Brian Moss and Sergeant Peter Venter are buried in the Beaumont le Roger Communal Cemetery, France. Huntley would be killed in action over Bremen, Germany on 14 September 1942.

enlivened by a burst of fire from one of the rear gunners and 'the speed with which they took cover did great credit to their instructor.' At a frontier post on the Swiss-German border an SS man in black uniform, black boots and black cap shook his fist at the low flying Lancasters. Crossing Lake Constance a German officer standing on the stern of one of the white ferry boat steamers fired his revolver at the bombers. Rodley could see him quite clearly, 'defending the ladies with his Luger against 48 Browning machine guns.' At Lake Ammer, the last turning point ten miles south of the target, an old bearded Bavarian standing on the shores of the lake took pot shots at them with a duck-gun. One of the gunners asked his pilot if he could 'tickle' him.

'No leave him alone' was the reply.

Accurate map reading, notably by Pilot Officer Desmond Ossiter Sands of Albany, Western Australia, Nettleton's navigator, with 32-year-old Flight Lieutenant McClure as extra man to read maps and bomb aim and 23-year-old Flying Officer Donald Stuart Reddy Hepburn RCAF flying with Sherwood, brought the two sections in each formation to their destination.

'Rising ground then forced us to fly a little higher" 'Jock' Penman recalled "and eventually we spotted our final turning point, a small lake. I had dropped back a little from Sherwood's section at this stage and mindful of the delay fuses on the bombs, made one orbit before turning to run in on the target. The river was a very good guide and the run in was exactly as shown in the scale model at briefing. A column of smoke beyond the target presumably came from Garwell's aircraft and it was soon joined by another…'

'The target was easily picked out" Brian Hallows recalled. "The situation of the factory in a fork made by the River Wertach and an Autobahn made it easy to identify and we bombed the hell out of it. The gunners were ready for us and it was as hot as hell for a few minutes.'

'We were belting at full throttle at about 100 ft towards the targets' recalled 'Rod' Rodley. 'I dropped the bombs along the side wall. We flashed across the target and won the other side to about 50 ft because flak was quite heavy. As we went away I could see light flak shells overtaking us, green balls flowing away on our right and hitting the ground ahead of us. Leaving the target I looked down at our leader's aircraft and saw that there was a little wisp of steam trailing back from it. The white steam turned to black smoke with fire in the wing. I was slightly above him. In the top of the Lancaster there was a little wooden hatch for getting out if you had to land at sea. I realised that this wooden hatch had burned away and I could look down into the fuselage. It looked like a blow lamp with the petrol swilling around the wings and the centre section, igniting the fuselage and the slipstream blowing it down. I asked our gunner to keep an eye on him. Suddenly he said, 'Oh God, skip', he's gone. He looks like a chrysanthemum of fire.'

As soon as 'Ding' Ifould the bomb aimer on 'U-Uncle' had let the bombs go, he heard Penman say: 'He's on fire!' Ifould looked over the side and saw flames streaming ten to fifteen feet behind Sherwood's Lancaster. 'Escaping vapour caught fire', recalled Penman 'and as he turned left on leaving the target with rising ground, the port wing struck the ground and the aircraft exploded in a ball of flame. (I was sure that no one had survived and said so, on return to Woodhall Spa but Mrs. Sherwood would not believe it and she proved to be right. I met Sherwood after the war. He had been thrown out of the aircraft, still strapped in his seat, up the hill and had been the sole survivor). As we ran in at 250 ft, tracer shells from light anti-aircraft guns on the roof of the factory produced a hail of fire and all aircraft were hit. Mycock's aircraft on the right received a shell in the front turret, which set fire to the hydraulic oil and in seconds the aircraft was a sheet of flame. It went into a climb and swinging left passed over my head with bomb doors open and finally burning from end to end was seen by my rear gunner to plunge into the ground.'

'Ding' Ifould learned later that 'P-Peter' had been hit and set on fire over a mile from the target when the 'vic' first got to the edge of the town. He and his crew could have done several things at this point. They could have climbed away to try to bail out or they could have jettisoned the bombs to make an immediate crash-landing. However, with the Lancaster's whole port wing on fire, they kept going. Mycock stayed on the bomb run until he knew his bombs had gone. Then his Lancaster reared up, passed over 'U-Uncle' and plunged to ground. There were no survivors. Flight Sergeant John Granville Donoghue the rear gunner was an American from Glen Head in New York State, serving in the RCAF. He left a widow, Cecelia. Warrant Officer Leonard Harrison MiD, Mycock's 30-year-old navigator, was an old school friend from his home town. He left a widow, Sheila Mary Harrison, of Chatham.

'Ding' Ifould dropped his bombs from about 300 ft and Penman dived again to the safer level of the housetops and cleared the town safely. Seconds later 'U-Uncle' was overtaken by 'Y-Yorker' the port Lancaster in the 'vic' piloted by Ernie Deverill. A shell had ripped the cowling of Penman's port inner engine and with only three engines still working, 'Yorker' seemed to overtake literally like a flash. As it passed 'Ding' Ifould caught a glimpse of a hole six ft by four ft wide in the metal of the fuselage. Deverill's aircraft had received a hit near the mid-upper turret and a fire had started. The oil recuperators had been punctured and burning oil was trickling down the fuselage into the well of the aircraft. Deverill told Pilot Officer Butler the navigator to put it out. Sergeant Irons the wireless operator left his set to help. Their bravery earned Edward Robert Butler a bar to his DFC and Ronald Percy Irons the DFM.

'By the time we arrived over the target area the element of surprise had completely gone' Irons recalled. "The German defences were very alert and were firing everything imaginable at us, including heavy gunfire. We were hit during the run-in to the target. Flak had hit the hydraulic pipes, which had put the gun turrets out of action and hydraulic oil had caught fire under the mid-upper turret. I left my position to help the gunner extinguish the fire. At the same time our starboard inner engine had been hit and was on fire. Having eventually extinguished the engine fire, we faced a long and hazardous journey back to base on just three engines.'

'Despite the distractions' recalls 'Jock' Penman, 'we held course and the front gunner did his best to reduce the opposition. My navigator was then passing instructions on the bombing run and finally called, 'Bombs gone.' We passed over the factory. I increased power and dived to ground level just as Deverill passed me with one engine feathered and the other three on full power although the u/s engine was restarted before reaching the coast. I called Deverill and he asked if I would cover his rear as his turrets were U/S. However, as my turrets were also U/S and Ifould had no wish to relinquish the lead, I told him to resume his position. Our attack had been close to the planned time of 2020 hours and as darkness took over we climbed to 20,000 ft for a straight run over Germany. It says a lot for Deverill's skill that he remained in formation until we reached the English coast and eventually landed at Woodhall Spa just before midnight.'³

Crews returned with graphic descriptions of close fighting; one rear-gunner spotted a German behind a machine gun on the roof and saw him collapse under his return fire. As the survivors turned westward, the light failed and the aircraft, led by 'Darkie' Hallows flew back without any opposition under cover of darkness. Hallows noted, 'The quintessence of loneliness is to be 500 miles inside Occupied Europe with one serviceable turret! Time, 2015 hours.'

Hallows returned safely and was one of eight officers to be awarded the DFC for his part in the raid. Later, Hallows wrote in his diary: 'One event sticks in my mind. Over half the bombs dropped failed to explode!' As Hallows said later, 'it was a bad way to spend an afternoon!'

The crews believed they had destroyed their target, as they had seen their bombs strike the plant. They had placed seventeen of the 1,000 lb bombs on the main tool shop, causing significant damage to the roof and walls, but the machine tools themselves were largely still functional. Five of the seventeen bombs dropped did not explode and although the others devastated four machine shops, only 3 per cent of the machine tools in the entire plant were wrecked. The 29% failure rate was somewhat higher than the usual 20% failure rate that was typical of UK produced high explosive bombs. U-boat engine production continued without apparent disruption.

Deverill delivered this report: 'Target attacked in formation. No cloud, from 400 ft bombs were dropped on target area. No.3 'P' was seen to catch fire in the air and crash. Very heavy flak and light predicted flak and S.A tracer. Aircraft caught fire on starboard side of fuselage and bomb bay but was extinguished by the efforts of the wireless operator and mid-upper gunner. Both mid and rear turrets u/s from target. Port outer

---

3. Ifould was awarded the Distinguished Flying Cross. He was afterwards appointed squadron navigation officer and in July promoted to acting flight lieutenant. In all, he completed one operation in a Manchester and seventeen in Lancasters before being posted to 109 Squadron in October and logging up forty-five ops on Mosquitoes. The most decorated navigator in the RAAF, he returned to Sydney in June 1944 and was transferred to the RAAF Reserve on 25 November. That year both his brothers were killed'; Elton Murray while serving with the RAAF and Frank Henry while serving with the Royal Australian Navy.

engine was u/s and feathered on leaving but was restarted before reaching coast. Formated again on No.1 'U' for protection and landed base.⁴

"'The trip home' said Patrick Dorehill 'was uneventful, thank goodness … Nettleton did a brisk circuit and down we came to be almost out of fuel. Golly, I can tell you I was glad to feel those wheels touch the grass.'

Squadron Leader Nettleton, who landed his badly damaged Lancaster at Squires Gate, Blackpool ten hours after leaving Waddington, was awarded the VC. 'Jock' Penman adds. 'Nettleton, it would appear, having increased speed to avoid fighters, bombed early and unable to cross Germany alone in daylight, turned back the way he had come. Due to navigational errors he eventually reached the Irish Channel and landed at Squires Gate. All surviving crews were grounded until a press conference, which I attended, was held at the Ministry of Information in London, when awards were announced.'

"Squadron Leader Nettleton's VC; immediately awarded – heaven knew, he had earned it" wrote 'Pip' Beck, a WAAF in Flying Control at Waddington…"but what of the others? Of our six aircraft Nettleton's was the only one to return; five did not – thirty-five men missing – thirty-five empty bunks, thirty-five empty places at table, shared between the Officers' and the Sergeants' Messes. Many of us wondered if it was worth it… At least Squadron Leader Sherwood and Warrant Officer Crum lived. But there were no medals for them."⁵

Tragically, the sacrifice of seven Lancasters and 49 young men on the Augsburg raid had been in vain. Oblique photographs taken some distance from the factory would not show much actual damage. But a careful reconnaissance was made later and photographs taken then showed that the whole roof of the main Diesel engine assembly shed, covering an area of 626 by 293 ft, was wrecked by high-explosive and fire. The largest hole is eighty feet across and, quite apart from the holes, the main structure of the roof is a wreck. The building has several storeys and

---

4. Squadron Leader Deverill DFC* DFM AFC was killed on 17 December 1943 near Gravely airfield Huntingdonshire attempting to land in fog conditions. He left a widow, Joyce Deverill of Hunstanton, Norfolk. He is buried at Docking (St. Mary) churchyard, Norfolk.
5. *A WAAF in Bomber Command* by Pip Beck (Goodall Publications 1989). The award of the Victoria Cross to Sherwood was 'strongly recommended' by Air Marshal 'Bomber' Harris but was not endorsed by the Air Ministry who substituted: 'To be recommended for DSO if later found to be alive'. His DSO was gazetted on 30th June 1942.

it is certain that the top storey is devastated. How many bombs penetrated farther down can only be guesswork, but there can be no doubt that, with such damage visible in the aerial photograph, the interior of the building and its machinery have been most severely damaged if not entirely wrecked. Repairs to the roof had begun, but had not got very far, when the photograph was taken.

Close beside the main assembly shop are two crank-grinding shops. They are smaller buildings, but of great importance both to the factory as a whole and in the production of Diesel engines. One of these shops was entirely demolished and the other, together with some sheds beside it, was severely damaged. Other buildings in the factory, including a shed 373 by 186 ft, the roof of which was destroyed, also incurred heavy damage.

Winston Churchill sent the following message to Arthur T. Harris, 'We must plainly regard the attack of the Lancasters on the U-boat engine factory at Augsburg as an outstanding achievement of the Royal Air Force. Undeterred by heavy losses at the outset, the bombers pierced in broad daylight into the heart of Germany and struck a vital point with deadly precision. Pray convey my thanks of His Majesty's Government to the officers and men who accomplished this memorable feat of arms in which no life was lost in vain.'

The Chief of the Air Staff sent the following message: "I would like 44 and 97 Squadrons to know the great importance I attach to this gallant and successful attack on the diesel engine factory at Augsburg. Please give my warmest congratulations and thanks." And from Air Marshal Arthur T. Harris: 'Convey to the crews of 44 and 97 Squadrons who took part in the Augsburg raid, the following: the resounding blow which has been struck at the enemy's submarine and tank building programme will echo round the world. The full effects of his submarine campaigns cannot be immediately apparent, but nevertheless they will be enormous. The gallant adventure penetrating deep into the heart of Germany in daylight and pressed home with outstanding determination in the face of bitter and unforeseen opposition takes its place amongst the most courageous operations of the war. It is moreover yet another fine example of effective cooperation with the other services by striking at the very sources of the enemy effort. The officers and men who took part, those who returned and those who fell, have indeed served their country well."

Chapter 2

'The Unwanted Hat Trick'

Edward A. 'Ted' Robbins was born on 26 February 1923. During his early boyhood the family moved from Wormley, Hertfordshire to Vancouver in Canada. His lifetime interest in flying began during many hours spent at Lulu Island aerodrome (now Vancouver International) and building many model aircraft. The family returned to England in 1938 and in August 1941 the slim, 18-year-old six-footer joined the RAF. In October he returned to America for flying training on the PT Stearman primary trainer at the Institute of Aeronautics at Tuscaloosa in Alabama. Ted completed his flying training at Gunter Field, Montgomery, Alabama flying BT-13 basic trainer and was then sent to Craig Field at Selma Alabama to fly the AT-6 Harvard advanced trainer. He gained his 'wings' on 3rd July 1942. That same October Ted was posted to 19 OTU Kinloss on the Moray Firth where he 'crewed up' with four other sergeants.[1]

Leslie 'Cal' Calvert, Ted Robbins' 30-year-old bomb aimer remembered his skipper as being 'a bit strait-laced'. 'Ted lived a narrow life. He didn't go out with girls and wouldn't listen to jokes about them. He didn't swear and never went into a pub; he would rather visit a church bazaar and have a cup of tea. He was a very close person. We didn't know much about him except where he came from. He never allowed pictures to be taken of the crew. Sometimes WAAFs wanted to look over the kite. When that happened Ted made himself scarce, but for all that he was a very good skipper and a lovely man.' Calvert had also trained as an air observer and was qualified to take over the controls if the pilot was killed or badly injured. At just 5 feet 4 inches in height and weighing less than 8 stone, he always kept his sleeves rolled down because he was embarrassed by his skinny arms. He was married and a trainee industrial welfare officer for a Manchester

---
1. See www.winthorpe.org.uk

dyestuffs manufacture. Leslie C. 'Les' Carpenter, the equally diminutive navigator, was a pleasant, straightforward and good-looking lad from Birmingham. George F. Calvert the wireless operator, no relation to the bomb aimer, was known as the 'Bolton Wanderer' because he supported his local football club, Bolton Wanderers. He also liked girls and beer. John N. 'Jack' Denton, the cheery 23-year-old rear gunner was from the Cornish village of Mousehole. He was always hungry and gobbled every meal as if it were his last.

After 25 hours of day flying on Whitley IVs, the crew training switched to bombing practice, cross-countries, photo flashes, loop bearings and QDMs at night. On 9 December 1942 Ted took Whitley BD222 'T-Tommy' off at 1645 hours for a night cross-country exercise. It was already pitch dark with strong winds and snow flurries adding to the unwelcome conditions. Soon after climb-out the port engine caught fire and had to be shut down.

'We had to fly over the Grampians and lost an engine about thirty miles from the airfield' Les Calvert recalled. 'It cut out and we flew like a brick, losing height all the time, so we turned round smartish. With the weather getting worse Ted fought to keep the aircraft in the air. Kinloss told us to fly around and await further instructions, but that wasn't possible. The only thing to do was crash-land on the grass at the airfield. Ted called me from the nose to sit beside him and we were only doing around 50 mph when he made a beautiful wheels-up landing about half-way across the airfield beside the dimly-lit runway with full flaps. We were carrying some small flash bombs which were stowed externally, port and starboard and they went off when we touched down. They gave off a flash and didn't set the aircraft on fire, but it was a bit frightening. The medics snatched us out and checked us all over to make sure we were still alive.'

The Whitley came to rest only a hose length away from the waiting fire tenders. The only slight injury was suffered by George Calvert, who bruised his knee on his table when the aircraft touched down. Less than three hours later the crew were airborne again in another Whitley for an hour's local flying!

On completion of the OTU course and after enjoying a spot of Christmas leave, the crew was posted to 1661 Conversion Unit at Winthorpe by Newark. There they collected a flight engineer, John Seedhouse and mid-upper gunner, Bertram M. 'Bert' Manley, 19, from Basildon, Essex. He was deeply attached to his bed, never interrupted sleep for breakfast and sometimes slept through lectures. Seedhouse, a Geordie, had joined the RAF two years before the war. Les Calvert described him as a 'shortish' fellow, not a glamorous type, quite unattractive but he could pull the 'birds'. He never went out with us because his first aim, no matter where we were, was women and it didn't matter how young or old they were. He was a loner and was off the first day at a new station when he would find a woman somewhere. He didn't go out with girls on the camp; she would have to be a civilian.'

On 9 January 1943 all seven crew members made their first flight together on Manchester R5838. After fifteen hours on type, they finally got to fly the Lancaster when on 8 February Ted flew R5547 on a 1½ hour local conversion trip. There followed two weeks of intensive familiarisation training, being airborne every day until, having completed 12½ hours they went off on their own without instructors for the first time, on a cross-country. On 21 February Ted Robbins took Lancaster R5892 off at 1120 hours for what was to be a simulated bombing raid on the Ruhr with eight sand-filled 1,000lb bombs in the bomb bay to give them the feel of the aircraft with an operational load. They had been briefed to drop two of the bombs at each of four practice ranges dotted round the English and Welsh coasts and which would be their en-route turning points; the first being in The Wash. From there they were to head south and alter course at Southwold, Suffolk to drop two more bombs on The Naze before turning southwest for their next objective, to Newquay, Cornwall. This was to be the longest leg of the flight. Finally, they were to turn north for Penrose on the Isle of Anglesey.

'We got to Southwold at 20,000 ft in sunshine' recalled Les Calvert. 'The sun was shining right in the nose of the aircraft. It was lovely, basking weather. I got my maps out and for the first time took off my Ingersoll watch and put it on the spur to the bomb sight. I forgot the watch later when we had to get out in a hurry, but for now I could see the time, I could see my maps and I could see the 10/10ths cloud below which hid the ground.'

Ted Robbins adds: 'We could already see the predicted cloud cover and now had to completely rely on the navigator. We would have no reason to have more visibility on the return to the base! When we were well on track, we decided to get something to eat and drink while we climbed to 20,000 ft. It was sunny and bright above 7,000 ft, perfect flying weather.'

'Everything was going well' adds George Calvert. 'Each member of the crew was thoroughly enjoying himself, getting to know each other. Two hours out and the wireless was working perfectly. I had tuned in to some pleasant dance music for relaxation and I was warm and comfortable. Then suddenly, everything changed. There was an almighty bang, as if the Lanc' had received a near miss from flak. It immediately plunged into a very steep spiral dive – virtually a spin – from which no-one thought it would recover. We'd never experienced anything like it before. Looking out through the window opposite my seat, I could see the starboard inner engine had lost its cowlings and the wing leading edge had disappeared right back to the wing root; a lot of skin from the top surface of the wing had also gone and it was fortunate none had hit the tailplane or fins. All four engines were going full blast.'

'We were over Reading, west of London at 19,000 ft," Ted Robbins said "when we heard the terrible blow. I was stunned by this sound, until Jack called me from the tail turret and asked what had happened and what the hell I was doing. I did not answer him, I had my hands full with the controls, which abruptly turned right and threatened to get us in a spin. When I got the plane under control, we had lost 6,000 ft. Now we were able to take in the damage. The entire plating and all piping between the fuselage and the right engine in the wing was gone. We were glad that the tailplane was not damaged. It was obvious that we could not finish our mission. George sent a SOS and tried to contact Air Sea Rescue. Les Carpenter reported that we had flown ten miles over the Channel and Leslie Calvert dropped the rest of the bomb load.'

'Between them the pilot and the flight engineer got the aircraft level again and continued on the same course" said Les Calvert. "We then considered what to do. Bailing out was an option, but we were running parallel to the coast and didn't want to land in the drink. The skipper said to me: 'Cal', I'm going to put the undercart down, will you check?'

'The undercarriage came down and I looked out of the port side. Everything was all right, the engines were ticking over; the wheels were fine. I peered out of the starboard side. Oh, my God! All the leading edge of the starboard wing had gone. All the cowlings of both motors had gone, so had part of the starboard wheel, the rubber and the mechanics. The rear gunner thought it was flak from British guns; he had seen bits and pieces flash past the tail. I believed that the aircraft had been sabotaged, probably by the IRA.'

'On Ted's instructions' continues George Calvert 'I contacted our home base to report what had happened. We were advised to head for Manston, which had a huge runway specially provided for such emergencies. Manston offered advice on speeds, etc and gave a heading to steer, but we were losing height and wouldn't make it; and in any case, it was found that the speed required to prevent the Lanc' stalling was too high, even for Manston. Closer examination of the starboard undercarriage revealed that not only was the tyre flat, but also one of the legs was bent upwards at a 30-degree angle. So Ted decided the only thing to do was put her down on the sea and this was relayed to both base and Manston Control before heading out towards the Channel at 160 mph, which was judged to be the best speed for maintaining a degree of stability and control. A lucky cloud break revealed the sea below and enabled us to jettison the remaining dummy bombs, thus relieving us of one problem. The crew set about ditching preparations. Our training was to be put to the test. I concentrated on sending a Mayday distress call, first ensuring the trailing aerial was out and the correct 'South Coast MF/DF' frequencies selected. There was a lot of Morse on the air, but it immediately ceased once the SOS had been tapped out and suddenly, I had the frequency all to myself, a tribute to the high standard of RAF discipline. It was comforting to receive a response and confirmation that bearings had been taken, with approximate position given. Our 'nav' had also worked out our position which I transmitted back to the ground station. The routines were operating splendidly. By this time the Lanc' was low over the sea and we were warned by our skipper to take up crash positions, so I clamped down the Morse key and crouched against the wing root bulkhead. We hit the sea with a great bump and water gushed rapidly into the fuselage, the aircraft soon coming to a halt.

'The crew was preparing for the emergency landing' continues Ted Robbins 'and I flew as low as possible at a speed of 150 mph. The blow on the water broke the tail off about the middle turret. John Seedhouse opened the hatch above my head. George Calvert pulled the cord on the inside of the fuselage to inflate the dinghy.'

'We pushed off the escape hatches on top of the fuselage. On my way out I picked up the Very pistol and a couple of cartridges and struggled through the hatch to see the dinghy inflating – a welcome sight. Ted was scrambling along the top of the cockpit and by this time the Lanc' was already beginning to sink nose down. The sea must have flooded in quickly through the damaged starboard wing. Six of us got into the dinghy relatively easily, stepping-in without even getting wet, but poor Bert Manley, the mid-upper gunner, was in the sea in full Irvin suit and flying gear, already very heavy. We had difficulty pulling him aboard and only managed it after a struggle. The aircraft was sinking rapidly and we cut loose the tethering rope. The Lanc' suddenly plunged nose down, followed by its tail section which had broken off on impact, the twin fins disappearing like the tail of a great whale. We estimated she had been on the surface 2½ minutes at most.

'There was a lot of jagged debris floating about, which we gingerly pushed away from our dinghy in case it got punctured. There was a gentle swell; at the top of which we could just see what we were fairly sure was our coastline on the horizon. After checking our Mae Wests and other bits and pieces of equipment and stores we started to paddle, somewhat hopefully. Already, I think we were grateful for all the dinghy drill we had done in training and which, at times, had seemed something of a chore. We felt quite elated and considered we had come out of it all rather well so far, with no injuries. We began to discuss what had happened, when we might be picked up and by whom. Being in mid-Channel, it could be either side we reckoned. At least we had the satisfaction of knowing someone knew we were down, our distress call having been positively acknowledged.

'We hadn't been in the dinghy long when suddenly three fighters looking very like FW 190s head-on were coming straight for us at low level in a 'vic'. Several of us prepared to dive into the sea, fearing a machine gun attack. The fighters banked steeply round us and with relief were identified as Typhoons, a type fairly new to us. Two of them maintained a protective orbit overhead while the third made off towards the coast, reappearing in our direction several times, obviously acting as a guide.

'Before long, we saw a column of smoke on the horizon and thought it a good idea to fire a Very cartridge, following which we watched the direction of the vessel. After a while, there was no doubt it was heading our way, so we fired our other cartridge. Someone was definitely coming for us! Soon we could see its outline and it grew into a small ship, which eventually stopped 100 yards or so away. Quite a few sailors lined the deck rail, their uniforms unfamiliar to our eyes and looking German-like.'

'We wondered if the ship was one of ours and if it had seen us' Les Calvert said. "When it turned towards us we could see it was flying an unusual tethered balloon, like a small barrage balloon with thin fins. It got within hailing distance and circled us. The sailors wore hats with tails at the back and a word at the front we couldn't read. We knew they weren't Royal Navy and wondered, a little fearfully, if they were German. They stood at the side of the ship looking at us. Some carried rifles.'

'A large man with a big red beard and who we guessed was the Captain' continues George Calvert 'bellowed out:

'Are you English?'

'Not knowing any other language, we could only answer: 'Yes, we're English'. What would be his reaction to this? The answer really surprised us: 'OK – We Dutch'.

'We paddled alongside and were hauled aboard along with our dinghy; and were now in very high spirits; lots of back slapping. We noticed the rifles being stowed away and I don't think German aircrew would have got quite the same reception. The ship was the Dutch mine-sweeper Zr. Ms. Rotterdam of the Klondyke Marine. 'Captain Red Beard' [Kapitein Gallandat de Hüet] told us he was heading for Portland Bill. Our luck was holding. Our wet clothes and flying boots were taken down to the engine room to dry. We agreed to let the crew have our flying boots and in exchange they gave us various items of footwear. I received a pair of gaily-painted wooden clogs. Until this moment, I had never seen skipper Ted smoke or touch alcohol; but here he was now with a cigarette in his mouth, mug of coffee in one hand, a cognac in the other, with the rest of us! The Dutch crew were a happy crowd; we didn't envy them their job.

'We heard the sound of aircraft engines and again wondered what we were in for. The plane was soon identified as an Air-Sea Rescue Walrus. It made a few low passes and headed for home, no doubt now as to our whereabouts and position being known. Shortly afterwards an ASR launch came speeding our way and its Captain tannoyed for details of how many airmen picked up and destination, etc. He decided not to transfer the seven of us, but to escort us into Portland. We were being so well looked after and were quite content to remain with the Dutch crew. On arrival at Portland and after thanks all round, we were taken to the Naval hospital for a bath and a short medical at which we were all pronounced fit and well. I was still wearing my newly-won clogs which were clumsy and awfully noisy. We were then taken to the ASR base at Warmwell for debriefing, this being where the launch had come from and where our SOS had been picked up. We also learned that an Army colonel, who had been walking along a cliff top, had actually seen us come down a few miles off St. Albans Head [on the Dorset coast five miles WSW from Swanage] and had immediately reported it. So, we had many things going in our favour on that day. We returned to Newark by train, a rather motley looking bunch and me in my clogs. At King's Cross station we were stopped by MPs who talked about putting us on a charge because of our appearance. We felt we had earned the week's survivors' leave which we were eventually given.'[2]

At the Court of Enquiry held into the loss of the aircraft, Ted Robbins and John Seedhouse were made to stand to attention for 1¾ hours while questions were fired at them. The theory eventually reached and accepted was that the starboard tyre had burst. The pilot and crew were highly praised for their actions. An extract from '5 Group News' later reported that 'a u/t pilot on his first Conversion Unit cross country ditched a Lancaster successfully. The conditions under which this ditching was made should give confidence to anyone who may have to repeat this performance. Although the aircraft ditched at 150 mph no ill effects were felt by any of the crew who carried out their dinghy drill religiously and with obvious success, having been instructed the previous day. The aircraft floated for 2½ minutes.' They had earned their Goldfish Club brevets.

In mid-March 1943 the crew joined 106 Squadron at Syerston near Newark in Nottinghamshire where Wing Commander Ronald Edward Baxter had taken over from Wing Commander John Searby as CO and Squadron Leader Peter Ward-Hunt was the 'A' Flight commander.

---

2. Quoted in *Lancaster At War: 3* by Mike Garbett and Brian Goulding (Ian Allan Ltd 1984).

Previously, the squadron had been commanded by Wing Commander Guy Gibson DSO DFC* before he left to form 617 Squadron. Ted Robbins, Les Calvert and Les Carpenter had each been promoted to Warrant Officer and the rest of the crew were now flight sergeants. On 22 March Ted Robbins flew the first of two '2nd dickey' flights as they were known, to afford the new pilot operational experience, prior to captaining his crew on a bombing sortie. In the early days heavy bombers carried second pilots or '2nd dickeys'. The idea was not so much as a reserve but rather to give the newcomer to the squadron a chance to learn his job thoroughly before tasking on a crew of his own. As time went on the number of '2nd dickey' trips was reduced to a couple, just to give the 'sprog driver' the inexperienced pilot, an idea of the 'score.' Flying Officer E. A. Edmonds was at the controls of ED360 on the trip to Sainte Nazaire. A few days later, on 26/27 March, Robbins was '2nd dickey' on W4842 ZN-H *Fema Dora* (RAF slang for 'Fuck-EM – All Dead-OR-Alive') for the raid on Duisburg. *Fema Dora* had been taken on charge on 106 Squadron in December 1942 and had been one of the aircraft flown by 25-year-old Flight Sergeant Lewis J. Burpee DFM RCAF when he was on the squadron. Burpee, recently married and from Ottawa, Ontario, received his commission shortly after Guy Gibson requested that he and his crew join him on 617 Squadron for the famous raids on the Ruhr dams on the night of 16/17 May. They were one of eight 'Dam Buster' crews that failed to return.[3]

On 28/29 March meanwhile, Ted Robbins and his crew flew their first op when they were one of 323 aircraft – 179 Wellingtons, 52 Halifaxes, fifty Lancasters, 35 Stirlings and seven Mosquitoes – that were detailed to bomb St. Nazaire. One Halifax and a Lancaster failed to return but Lancaster W4156 and Ted Robbins' crew returned safely. W4156 went missing shortly afterwards, on 8/9 April, when 23-year-old Pilot Officer John Lawrence Irvine's crew failed to return from the raid on Duisburg. There were no survivors. Irvine left a widow, Jessie Dalrymple Irvine

---

3. As well as Burpee, Gibson had personally selected Flight Lieutenant Dave J. Shannon DFC RAAF and crew and Flight Lieutenant John Vere 'Hoppy' Hopgood DFC* and crew on 106 Squadron. 53 men were killed and three were captured. Burpee's and Hopgood's crews were among those lost. Lancaster ED865 AJ-S and Burpee's crew were shot down at 0200 hours immediately south of Gilze-Rijen on the way to the target while following the Wilhelmina Canal between Gilze-Rijen and the flak defences at Eindhoven.

It was Berlin on 29/30 March when weather conditions were difficult, with icing and inaccurately forecast winds.[4] Once more *Fema Dora* returned safely though 21 other bombers did not. Within two months Ted Robbins' crew had completed a total of sixteen night ops. Some were to distant targets such as Pilzeň in Czechoslovakia on 16/17 April – a long 'flog' of almost ten hours on *Fema Dora*. It was the night that 'Bert' Manley the mid-upper gunner nearly distinguished himself by shooting down a British bomber, as Les Calvert recalled. 'The weather was terrible, very heavy cloud and we went across France at 5,000 ft. We had to climb to 20,000 ft after crossing the Rhine at Strasbourg. 'Bert' was first through the cloud and he saw an aircraft which he fired at. Luckily he missed. It was a Halifax, which fortunately did not reply. Ted Robbins told him mildly: 'We'll have less of that'.'

An even longer 'flog' was to La Spezia on 18/19 April when *Fema Dora* made a round trip lasting ten hours and ten minutes. The aircraft received a flak hole in the bomb aimer's compartment but made it back safely. The crew rode their luck on Stettin and Duisburg: coned by searchlight for eight minutes over Stettin; an engine knocked out over Duisburg; but always, Ted and the Lanc' brought them home. They were stood down on Tuesday the 27th when 160 aircraft carried out the biggest mine laying operation of the war so far with 458 mines being laid off the Biscay and Brittany ports and in the Frisian Islands. One Lancaster failed to return. The threefold purpose behind the mine laying trips was to seriously impede the flow of men and materials passing through Baltic ports to the Russian Front; to dislocate U-boat training, which was carried out extensively in the Baltic; and to interrupt and delay the highly important shipments going into Germany from Sweden and Norway via the Baltic.

On Wednesday 28/29 April Ted Robbins and his crew were one of 68 Lancasters in a force of just over 200 aircraft detailed to drop mines off Heligoland, in the River Elbe and in the Great and Little Belts. Low cloud over the German and Danish coasts forced the mine-laying to fly low in order to establish their positions before laying their mines and

---

4. *The Bomber Command War Diaries: An Operational reference book 1939–1945* by Martin Middlebrook and Chris Everitt (Midland 1985).

much German light flak activity was seen.⁵ Ted Robbins' Lancaster was caught in a cone of searchlights as he passed over one of the Frisian Islands. 'There seemed only one thing to do' he recalled – 'shoot them out. My gunners fired right down a beam and I saw the bullets hitting the projector. The searchlight went out. My front gunner sprayed another light ahead of us and the beam twirled away as if the crew had abandoned their post and were running for cover.' Twenty-two aircraft – seven Lancasters, seven Stirlings, six Wellingtons and two Halifaxes were lost. This was the heaviest loss of aircraft while mine laying in the war, but the number of mines laid was the highest in one night.

On the night of 30 April/1 May Robbins and his crew rounded off the month with a raid on Essen and followed this raid with visits to Dortmund, Duisburg, Pilzeň and Dortmund again and Düsseldorf. On the trip to Dortmund, on the night of 23/24 May, they made an early return after 2.55 hours had elapsed after the port inner engine failed while climbing out from Syerston. That same night Wing Commander Ronnie Baxter was forced to abandon the operation when his port outer engine failed and he jettisoned the 4,000lb bomb but brought his incendiaries back. On the night of 25/26 May *Fema Dora*'s crew hauled a 4,000lb 'cookie' and 125 (4lb) SBCs (small bomb containers) to Düsseldorf. There were two layers of 7/10th cloud and haze over the target and the Path Finders experienced great difficulty in marking it. No yellow Target Indicators (TIs) were seen the target area, which was located by a cluster of Green TIs and Les Calvert got the bombs away at 20,000 ft at 0209 hours but no results were observed. The attack failed, probably due to the PFF Red TIs being exceptionally scattered. Very few fires were seen and these were over a very wide area. Heavy flak was predicted and accurate and 27 aircraft were shot down.

Two nights later, on 27/28 May, when 518 aircraft – 274 Lancasters, 151 Halifaxes, 81 Wellingtons and twelve Mosquitoes – were detailed to raid Krupps of Essen, the most hotly defended of all targets; fate took a hand. Bert Manley reported sick with a throat infection but Ted Robbins was keen to go and he asked Wing Commander 'Ronnie' Baxter for a spare mid-upper. The Wing Commander was not flying that night, so he

---

5. *The Bomber Command War Diaries: An Operational reference book 1939–1945* by Martin Middlebrook and Chris Everitt (Midland 1985).

offered the services of his own regular mid-upper, Sergeant Howard John Taylor, adding, 'he's a nice boy from Birmingham, only nineteen; make sure you take great care of him.' The offer was accepted but 'Spud' as he was more familiarly known came in for some leg-pulling because some of Ted's crew thought he was 'disgracefully upper-class'.[6]

Ted Robbins took *Fema Dora* off from Syerston at 2150 hours in fine weather for their third trip to Essen. In the bomb bay was a 4,000lb 'cookie' and twelve SBC (4lb) small bomb containers. 'By the time we were on course' recalled Robbins 'it was already pretty dark and as far as we could see we were the only aircraft in the sky, even though we knew that at least there had to be 400 to 500 other aircraft in the air. Over the North Sea, as usual we tested the guns while still climbing.' Their route out was via Egmond and the return would be made via Bunschoten and landfall would be made at Cromer on the Norfolk coast. *Fema Dora* arrived at the target, which was cloud covered, at 0042 hours at 20,000 ft. Flak over the target was intense and sky marking had to be used because of the cloudy conditions. The main bombing was scattered, with many aircraft undershooting. 'At one point' continues Robbins, 'we could see a purple-red glow in the distance. We knew that a bomber had been hit and exploded.[7] When we approached the first flak area we were at an altitude of 22,000 ft. We were on the marker flares and 'Cal' took over. He gave me the heading and then he said 'Bombs gone!'

'I finished the bombing run 'smartish' – probably quicker than I should have' Les Calvert recalled.

Ted Robbins could just see their bomb load explode on the ground and simultaneously took a picture. 'Immediately afterwards we were hit. I lost control but finally, after much effort I got things under control, but we were certainly 5,000 ft lower. I called all the crew members one by one and each one said that they were okay. Then everything happened very quickly. George reported that the two left engines were on fire. Johnnie Seedhouse feathered the propellers. I tried to dive to extinguish the flames but unfortunately, this did not work. Johnnie fought the

---

6. On 6 May 1945 Bert Manley, who had joined 467 Squadron RAAF on 9th July 1943 and completed 29 operations by 20 March 1944, received his commission. He was de-mobbed in April 1946. *Ted Robbins Bommenwerper Piloot* by Dick Breedijk.
7. Altogether, 23 aircraft – 11 Halifaxes, six Lancasters, five Wellingtons and a Mosquito FTR.

flames with a fire extinguisher. Fortunately there was no damage to the wing or fuel tanks. Luckily, two engines on the right side were still running and nobody was hurt. Then we were hit again and I asked the crew if there was any damage. George reported several holes in the floor. Jack reported that the rear part of the aircraft seemed fine. But I felt the aircraft shake. After releasing the bombs 'Cal' had flipped the switch to close the bomb doors but we were hit and the bomb doors were wide open. I was just over the shock of this information when Johnnie called 'right inner engine on fire.' I told him to feather the prop. I would rather not use the fire extinguisher because I wanted the opportunity to restart the engine. I asked Les to give me the shortest course to the coast. Les gave me the route (not the shortest) and I went on track. I gave orders to 'Cal', Jack and 'Spud' too, because we could fall easy prey to the night fighters. I did not know how long one engine would hold out under these conditions and continuing to fly in a straight line was difficult on full right trim.'

The 'Viermot' ('four engined bomber') was claimed destroyed by Leutnant Walter Schön of 3/NJG1 for his fifth victory but they had been hit by flak first. (Schön was killed in action on 20 October. He was credited with a total of nine victories). Ted Robbins fought to pull the Lancaster out of its plunge, eventually doing so at 5,000 ft. The fire extinguishers doused the flames, there were no injuries to the crew and there seemed a chance they could make it. The Lancaster was shaking severely and the bomb doors were jammed open, creating enormous drag and turbulence. 'You didn't normally go out of the target the way you went in because the Germans were waiting for you' recalled Les Calvert. 'But it would mean going further if we did a dog leg out. We decided to come back the shortest way and run the gauntlet. We didn't know if we'd lost any fuel because there were holes all over the place. We were losing height all the time. The skipper was having great difficulty in keeping on a course because we were crabbing, being pulled to starboard by the two starboard engines.

'Finally', continues Ted Robbins 'I knew we would not make it and I switched on the intercom and said: 'This is a good chance for anybody to get out if they want to while we still have enough height. I'll hold the kite steady for them.' I asked each individual crew member if they wanted to jump. Everyone decided to stay and see if we could reach the North Sea. All except 'Spud' Taylor had been rescued when we had ditched before.'

'We wanted to get back if we could and discussed the options' said Les Calvert. 'Our replacement mid-upper gunner, 'Spud' Taylor came on and said he would go and Ted told him to leave through the back door. A few minutes after Howard left his turret he plugged into the intercom at the rear door. Howard had his parachute on and the door open, but he was apparently torn by terrible indecision.'

Taylor asked 'what are you fellows going to do?'

George Calvert replied in his broad Lancashire accent: 'We're going to get down in t' drink.

'In the drink?' Taylor enquired, nervously.

'Ah, if we don't mek it home we'll get down in t' English Channel somewhere.'

'Isn't that dangerous?'

'Nay, lad. We've been in t' drink before. We've done it before we can do it again.'

Taylor cheered up and said: 'Oh, I'll come with you then. I'll get back in the turret.'

'We decided therefore to throw all excess equipment out of the aircraft' continues Ted Robbins. 'George, 'Cal' and Johnnie tore out machine guns, ammunition, ladder, bomb aiming equipment; anything we did not need.'

Les Calvert adds: 'Then we came to the music hall bit, throwing things out to help us maintain altitude. I uncoupled the bomb sight, a great big thing and passed it back to the flight engineer who gave it to George Calvert who threw it out the back door. Everything we did not need, including the ladder, was thrown out and George shut the door, but we were still losing height, down to about 2,000 ft and there was flak all over the place, but the marvellous thing was, none of us had been scratched.'

Only the starboard-outer was left working at full bore and even with full trim, it took both feet on the right rudder for Ted Robbins to counter the pull of the one engine. 'The plane lost altitude and we needed more speed so I un-feathered the propeller on the starboard inner, gave the engine fuel and tried to re-start it. The engine started immediately but began to burn again so I switched everything off, again without using the extinguisher. How long could we still fly on one engine?' It began dragging the Lancaster on a gently curving track directly over the middle of Amsterdam, where they became a sitting target for what seemed like every searchlight and AA gun in the city.

"The Germans let us get over the city centre before switching everything on" Les Calvert recalled. "We were coned by hundreds of searchlights; it was like daylight. Then their guns opened up. By now Ted had called me back into the crash position to release the escape hatch when necessary. We were hit by flak and I saw a row of holes suddenly appear in the port side of the fuselage and the light shining through them but, extraordinarily, no one was hit. It was a question of how far we could go. The skipper did his best to keep the aircraft straight and level and avoid going into a stall, but the single engine pulled him off course. I stood behind the main spar on intercom, chatting to help boost his confidence. Ted was a bit unhappy. I had a little joke with him and said he could do it. I reminded him of his wonderful landing when we ditched, but he was worried and said he couldn't really see where he was going.'

"Les kept me informed about the distance to the coast' continues Ted Robbins. 'We were at 3,000 ft above the flatlands of the Netherlands. The North Sea was so close yet so far away. George tried to contact base and give them our position and where we thought we would make our emergency landing in the sea but he failed to get through. [The trailing aerial had been shot away so there was no chance of another SOS signal, even if they were to ditch]. Les said that there was a stretch of water southwest of Amsterdam and if we did not make the sea it would make a good crash site. But searchlights and heavy shooting began and I told my crew to take up their crash positions. This was around 1:51 am. We could not shoot back because we had removed all weapons except for the upper turret. Finally the shooting stopped and the entire crew had survived! The aircraft was like a colander. I'd had a number of bullets go through the nose.

'We were very low and I decided to put the landing lights on. I was hoping to see something. It turned out well. After two or three minutes I thought I saw the waters of the North Sea or the Zuider Zee. I did not care what it was. I had sought and found water! (It was Lakerpolder near Sassenheim, five kilometres north of Leiden in Holland). I cried; 'Preparing for ditching'. Looming in front of me I saw a church steeple, a sail of a windmill and then another church tower.'

'The aircraft was difficult to control' continues Ted Robbins. 'Almost stalling I had to swerve to the right and give full throttle as I jinked round the windmill and the church steeple. I cried: 'Hang on, we're touching down'.

'The left wing went down'. I saw more water. Down we went. We hit, skipped a couple of times on the surface and water gushed inside the fuselage as we bumped along a low dyke left of the windmill before we flopped on to a muddy island and slithered across a field of barley. The aircraft slid to the right and I gave the starboard engine more power. This caused a sharp turn to the left before we came to a halt in peaceful silence. It was a miracle!'

None was more relieved that Les Calvert, who had looked out and saw the slowly-revolving sails of the windmill above them. 'The skipper had pulled off yet another copybook belly landing. The time was 0159 hours. Calvert concludes: 'It was as quiet and lovely a landing as could be, very soft. It was amazing that the kite didn't turn over; we were not even thrown about. It was as if we had landed on wheels, brilliant flying, probably unique. No one was hurt. Ted turned off the engine and we all got out.'

The crew decided to destroy the aircraft but to no avail. When the destructive devices failed to work, they fired off the Very pistols – again to no avail. In their haste to get away they forgot the pigeons, so the bomb aimer and the wireless operator returned to the aircraft. Les Calvert then wrote a message and put it in one of the leg rings before releasing the two birds and never did find out if they delivered it! He said: 'We wondered what to do because there was no vegetation and no buildings. It was a question of where to hide. I had a map or two but none of that particular area. I fancied my chances, having been in the Boy Scouts, but didn't know we were on an island. I left the aircraft to have a look round.' He wandered off through the churned-up field. After the war the farmer would make a claim against the RAF for the loss of his crops. George Calvert and Jack Denton joined the bomb aimer and the three men, still light-headed from the elation of being alive, set off along the edge of the water which they believed to be the Dutch coast.

'After about twenty minutes' Calvert recalled, 'we heard the 'plop-plop-plop' of a little motorised punt which had two Dutchmen in it. They looked at us and hove to. We made them understand that we were RAF airmen and they gave us a jam sandwich each, which was very kind of them. We all got into the boat and, after a while, stopped near a lot of residential caravans. In one were three girls of about twenty, one of whom spoke beautiful English. She gave us a cognac each and said they daren't take us inside otherwise the Germans would shoot them. We got back into the punt and eventually came to a small landing stage near one or two buildings. One Dutchman went off and bugger me; he came back with a German soldier. That was the end of our freedom. The other four chaps were picked up a couple of days later. We were taken to Amsterdam jail where, bollock naked, we were minutely inspected, every aperture examined. Even our stools were taken away for examination in case we had eaten coded information written on rice paper. We were taken into separate cells. Mine had no window, there was a po standing in straw on the floor and an arc light fixed to the ceiling. After a few days a German officer ordered me to identify one of my crew. I thought: 'Bloody hell! One of them's been shot. He took me down the corridor to where a guard was pointing a machine gun into a cell. Inside was a most dishevelled Ted Robbins, with a dirty face, no badges of rank and his epaulettes hanging down. He'd been given a bad time. He was wearing Wellington boots and I wondered where he'd got them from. The officer ranted at Ted, arms flailing, then turned to me and said: 'Herr Calvert, do you recognise this man?' I looked straight ahead and just gave him my name, rank and number. The officer bawled triumphantly at Ted: 'There's a man who knows his Geneva Convention.'

'A few minutes after I was marched back to my cell the officer reappeared and said: 'Herr Calvert, since you know your Geneva Convention we will let you stay with your brother George.'

'I was marched along to 'brother' George, who was in a black cell, one without a light, it was terrible. We sat together, whispering in case the cell was bugged. We were there a couple of days.'

Amsterdam railway station was cordoned off by scores of Germans when Les and George Calvert were among eighty prisoners waiting to be taken to Frankfurt for interrogation. 'A small Dutchman, not noticing the cordon, ran through it to catch his train' continues Les Calvert. "A feldwebel told a guard to stop him. 'The guard hit the poor man hard on the head with the butt of his rifle. He fell face down with blood pouring from a great gash in his skull. No one was allowed to help him. We knew then what we could expect from the Germans.'

The crew endured the rest of the war behind the wire at Stalag Luft III, Stalag Luft VI Heydekrug and Stalag XIB Fallingbostel near Belsen. In 1945 they were flown home. 'It was marvellous looking through a window of the Dakota and seeing traffic driving on the left-hand side of the road' Les Calvert recalled. 'We owe so much to Ted Robbins. If it wasn't for him and his amazing skills as a pilot I wouldn't be here.'[8]

---

8. See *Lancaster At War: 3* by Mike Garbett and Brian Goulding (Ian Allan Ltd 1984) and *Ted Robbins; Bommenwerper Piloot* by Dick Breedijk (2011).

Chapter 3

'Sheff'

At the end of our course at No.1 Lancaster Finishing School at Hemswell early in 1944 we went on leave and reported back on the morning of 6 June. We didn't know at the time that the "D-Day" invasion had started but we nobody said anything to us. We got into a crew bus and made a short journey to Ludford Magna near Louth where we'd been posted to 101 Squadron in 1 Group. We didn't know what this special duty meant, but we had an inkling.

At 16 Jack Morley had worked on the afternoon shift at the casting shop at Gladwin Brothers cutlery firm on Rockingham Street in Sheffield during the blitz on the nights of 12/13 and 14/15 December 1940. The first raid in Schmelztiegel ("Crucible") lasted nearly nine hours and was concentrated mainly on the centre, north-west and south-east of the city. Over 660 people were killed, 1,500 injured and 40,000 made homeless and 3,000 homes were demolished with a further 3,000 badly damaged. A total of 78,000 homes received damage. 'Jack' Morley turned down Angel Street where Cockayne's was blazing fiercely. "In the middle of the road was a pair of gauntlets, clutching a hosepipe. There was no body attached to the gloves. What a sight." Gladwins was bombed and there was no more work, so for a week or so Jack Morley and three other lads were made 'Messenger Boys' in the Home Guard. When he turned 18 'Sheff' enlisted in the RAF. He passed the aircrew selection board and came top of his wireless operator/gunner course. Don Dale, a little cockney, and 26-year-old 'Ginger' Congerton', tall and of course ginger haired, who claimed to be the two best gunners in the RAF and were top of their course invited him to join them. They talked two Welsh flying officers, 34-year-old G. A. John Arthur and Dai Jones into joining them. In April 1944, before being posted to 1667 Heavy Conversion Unit at Sandtoft to convert onto the Halifax, they had picked up a flight engineer and Flying Officer George Harris had become their skipper. He was sufficiently conversant with the aircraft and his crew was asked, "Are you ready for posting away?" 'Yes' they said but were told that they would not be going on a squadron yet because they were joining 101 Lancaster squadron, which was a great relief for them all – they didn't want to be on Halifaxes.

'Ludford Magna [or 'Mudford Magma', as it was known because it was still marshy although built on one of the highest stretches of the Lincolnshire Wolds] was strangely quiet. There were people standing around and walking about and my old pal Art Wallace was there to greet me. He'd got there a couple of days before and had seen our names on the list of arrivals. His crew hadn't got their special operator yet – neither had we until later. It was a strange morning. We were informed about what had been happening via a Tannoy announcement and were told that 24 aircraft on 101 Squadron had flown all night without bombs, just jamming the German night fighter wavelengths and dropping bundles of 'Window' to fool the Germans. Three Lancasters returned with either crew sickness or faulty aircraft and one crew that ditched off Beachy Head[1] was picked up by a destroyer, but the remainder flew around at 25,000 feet on a rectangular course. 101 had only six fighter contacts and only one was an action one. Other aircraft jammed the wavelengths of the defences; so all in all, it was a complete success. All the squadrons were congratulated for saving the 'D-Day' landings from being a massacre.[2]

'For a few days we wandered about, learning all about the place and the surrounding countryside. [J. C.] Colin York – a Lancashire lad – was appointed as our 'special operator'. Me being a Yorkshireman he was one of our traditional enemies and we had much banter about this. The navigator would join in and say: 'The wireless operator is the only man in this crew with a trade' and Colin would retort with: 'Wireless operator isn't a trade, it's a bloody disease.' Colin York had two receivers and a transmitter sited in the fuselage aft of the bulkhead between the wireless operator and the mid-upper gunner. The 'special operator' had

---

1. Lancaster I LL833 SE O flown by Pilot Officer M. J. Steele RNZAF.
2. Operations 'Taxable' and 'Glimmer', both devised by Wing Commander E. I. Dickie created 'Phantom Fleets' on enemy radar screens. 'Taxable' involved 16 Lancasters on 617 Squadron and was a joint RN/ operation aimed at making the Germans believe that an invasion force was attacking the French coast between Dieppe and Cap d'Antifer. Attacks on enemy radar installations had all but destroyed their effectiveness, but care had been taken to leave enough operational to allow the Germans to deceive themselves that their radars were showing an invasion fleet.

no windows to look out of like I did. He operated a receiver that had a white strobe, which moved backwards and forwards every fifteen seconds. When it lit up, it indicated that it was onto a German station. He would lock onto it, tune the transmitter and jam the frequency with a sort of 'wigwag; noise. He could do this three times. He had three transmitters.

'We gradually became acclimatised with Colin. Then came 29 June, our first operation, in W4967 'P-Peter', a 1942 vintage aircraft, to Siracourt, a big V-1 launch site [three miles WSW of Ste-Pol-sur-Ternoise] in the Pas-de-Calais. For about three weeks the flying bombs had been creating havoc in the south of England and it was decided that we would try to take the launch sites out. There must have been dozens of launch pads at Siracourt. If we gained height, there was nowhere safe; in fact the sky was full of aircraft to-ing and fro-ing from Siracourt.[3] As we approached the target, each aircraft had very accurate predictor flak coming at it. When we began our bombing run, predictors appeared at our height when… 'left-right-steady-bombs gone' the aircraft lifted upwards and George turned to port and the predictors were bursting where we would have been a second later. On the whole, it was a very good bombing, but the V1s continued coming.[4]

'On the night of 30 June/1 July we were scheduled to bomb the first of many marshalling yards, at Vierzon. The idea was to deny the enemy the use of any reinforcements that might be coming up by rail. We took off in LM462 'V squared' and gradually made height and headed towards Paris. Our mid-upper gunner said, "Jack', just have a look out.' There were two little handles, one each side of the window and I dared not sit down. I was absolutely scared to death for as far as I could see was a wall of flak and we were heading straight into it. Dai did his usual 'left, left, right, right' stuff and we flew through the flak and nothing burst near to

---

3. 286 Lancasters and 19 Mosquitoes of 1, 5 and 8 Groups attacked two flying-bomb launching sites and a store. There was partial cloud cover over all the targets and some bombing was accurate but some was scattered. On the raid on Siracourt three Lancasters, including that of the Master Bomber (and two Mosquitoes) were lost. *The Bomber Command War Diaries: An operational reference book 1939–1945* by Martin Middlebrook and Chris Everitt (Midland 1985, 1990, 1995).
4. Lancaster I W4967 SR-P had only been allocated to 101 Squadron in June 1944. It was lost on the raid on Homburg on 20/21 July. Pilot Officer Sydney Emmington Fern Higgs Smith RCAF and six of his crew were killed. Only one crew member survived.

us although it looked so fierce. With bombs gone, we turned for home. Was I relieved when we got away from that flak![5]

'The 4th of July was another marshalling yard target; this time Orléans, in LM472 SR-T. We took off at 2245 and crossed the coast at Dieppe. There seemed to be fighter contacts all around, which was only natural. Down to Orléans where we found the target and we had a good bombing run, turned for home. There were still many fighter contacts, but fortunately, none with us. We managed to steer clear of them right up to the coast at Dieppe and then the attacks ceased.[6] Then, on 5/6 July, we went to another marshalling yard at Dijon, but this time we used an aircraft that we came to think of as our own: LL756 SR-Z, taking off at 2225. This was a very good operation and we made a good attack. We didn't encounter any fighters on the flight which was five hours, 25 minutes.[7]

'These three marshalling yards had been a good introduction to night time bombing. We were quite pleased. Once more another marshalling yard (we were in LM462); at a place called Bon Ravinie. It wasn't too far inland, but our route was to take us across the coast at Dieppe again. We'd all the usual fighter contacts and some bombers going down. As we approached Dijon, we turned east as if going towards Stuttgart and then suddenly turned north towards to Bon Ravinie. Most aircraft did not manage to find a target and we were among them, yet the duration of our flight was the longest operation we'd flown: nine hours and five minutes in total. Many aircraft were lost whilst looking for the target. We were told to divert to an American aerodrome at Wigsley, just outside Lincoln

---

5. 118 Lancasters of I Group attacked the rail yards at this small town south of Orléans and bombed with great accuracy, a success for I Group's own marking flight. Fourteen Lancasters were lost. *The Bomber Command War Diaries: An operational reference book 1939–1945* by Martin Middlebrook and Chris Everitt (Midland 1985, 1990, 1995). LM462 SR-V² was lost on Stuttgart on 28/29 July. Pilot Officer Peter Joseph Hyland, the 21-year-old pilot, who was from Argentina and all his crew were killed.
6. 282 Lancasters and five Mosquitoes of 1, 6 and 8 Groups attacked rail yards at Orléans and Villeneuve. Both targets were accurately bombed. Fourteen Lancasters were lost; eleven on Villeneuve and three on Orléans. *The Bomber Command War Diaries: An operational reference book 1939–1945* by Martin Middlebrook and Chris Everitt (Midland 1985, 1990, 1995).
7. The main railway area was heavily bombed by 154 Lancasters of I Group. No aircraft were lost. *The Bomber Command War Diaries: An operational reference book 1939–1945* by Martin Middlebrook and Chris Everitt (Midland 1985, 1990, 1995).

because there was such a heavy mist and so many aircraft had gone to Ludford that there was no room for us to land. We stayed only until the next day when most of the aircraft had gone from Ludford.

'Next was a daylight raid to Lordeville; not a very long trip, this time in LM478 SR-A. We did a good job, hardly had any trouble and got back safely. On 18 July we took off for our first trip into Germany to the Ruhr – the Happy Valley – as we called it. There were all sorts of things in the Ruhr Valley, so we just bombed what we were told to bomb.[8] We took off at 2230. It was quite a harrowing trip of five hours' duration in our old faithful LL756 SR-Z and we were glad to get back home safe. Now a trip to 'Happy Valley'; one storage depot and two factories. Four hours duration this one. We took off at 2230 hours. Once again, a memorable trip; no serious damage to us and we got safely back. According to our intelligence officer we'd done a good piece of bombing.

On 23 July, in LL756-SL-Z, we were ordered to Kiel. We were told that after the First World War, following attacks at Kiel, a rebellion had been caused which helped shorten the end of the war. We were ordered to bomb the old town and we set many fires going, in the hope of causing a rebellion this time, but it didn't happen. To get to Kiel, we made many diversions without going inland, bombed and then came safely home. We had much flak, but no fighter contact.[9] This was to be my last trip before going home on leave, potentially to get married, but it didn't happen because I was asked to do another trip before I went on leave.

'On 24 July the trip of about eight hours' duration was to Stuttgart. We took off and all the ABC operators were working fully and we had a more or less, uneventful trip. We didn't make any contacts at all. We bombed the tool factory as ordered and then set off back.[10] Shortly after leaving

---

8. On 18/19 July 194 aircraft of 1, 6 and 8 Groups attacked the synthetic-oil plant at Wesseling 9 miles south of Cologne and 157 Lancasters and 13 Mosquitoes of 1 and 8 Groups raided the Hydriewerk synthetic-oil plant at Scholven-Buer. Four Lancasters FTR.
9. In the first major raid on a German city for two months, 629 aircraft (519 Lancasters, 100 Halifaxes and ten Mosquitoes) were dispatched. The elaborate deception and RCM operations combined with the surprise return to a German target completely confused the Nachtjagd and only four aircraft – all Lancasters – were shot down. *The Bomber Command War Diaries: An operational reference book 1939–1945* by Martin Middlebrook and Chris Everitt (Midland 1985, 1990, 1995).
10. This was the first of three heavy raids (by 461 Lancasters and 153 Halifaxes) on Stuttgart in five nights which caused serio7us damage to the central districts of the

Stuttgart – the special operators were still operating – we came into a layer between two clouds. Our skipper said to the navigator, 'Give me any change of course, but do not change height or direction anywhere.' We never saw another aircraft all the way back to England. A fine piece of flying by George. Although we didn't like him personally, he was a good pilot. After we landed, I had breakfast then a couple of hours' sleep and then went to get my leave pass, to go home and get married. I was supposed to have got married on 29 July. As it happened, this had been called off, which I didn't know until I arrived home on the afternoon of the 25th. Of course I enjoyed the leave, but I didn't like the idea of having to wait to get married. I was told that it'd been changed to 31 July and then to 1 August and eventually, 3 August. We got married and stood on the steps of the Wicker Congregational Church on Gower Street having photographs taken when the telegram lad came up with two telegrams. The first one was from my granddad, wishing us 'Every Happiness'. The other one was from the squadron: 'Be back on the camp by 8 o'clock on Saturday morning for operations.'

'Well, I didn't want to go back; I'd just got married.

'On Friday, the fourth, Beattie said to me, 'Look 'Jack', you've got to go back. We can't have a honeymoon. I'll go to Ivy's farm outside Boston. I'll go as far as Louth with yer and then you make your way to Ludford.'

'This was agreed finally, so Beattie went on her own honeymoon. (I learnt later that she'd been potato picking that week). I got a lift to Ludford and arrived just before 8 o'clock, just in time to meet the boys who were going to briefing for a daylight raid [on 5 August], a special one, to a place called Blaye [on the River Gironde] on the Bay of Biscay. I didn't like the idea, but I was on bombers and I had to do my duty. After briefing, we went for a meal and shortly after that, we took off. We flew down towards Land's End, flying in formation, coming down to sea level and winding out the trailing aerial before reaching the coastline. We went further until the meters flickered in front of me, showing that the aerial was earthed. We were flying at exactly fifty feet, which was fifty feet less than the aircraft altimeter showed. In this way, we continued

---

city, which, being situated in a series of narrow valleys, had eluded Bomber Command for several years. Seventeen Lancasters and four Halifaxes FTR. *The Bomber Command War Diaries: An operational reference book 1939–1945* by Martin Middlebrook and Chris Everitt (Midland 1985, 1990, 1995).

on a course fifty miles from the French coast and flying at sea level. We couldn't see the horizon. Every time the needle flickered, I said, 'Now Skipper' and he pulled the nose up a little; so we were flying at more or less fifty feet. We wouldn't fly into the sea because I let the skipper know every time the needles touched.

'Suddenly, the navigator gave a change of course and we went west towards Blaye. We climbed to 3,000 feet and then we dropped our bombs. The concentration was on oil storage tanks at an oil depot. We were among the first to bomb.[11] We reeled in the trailing aerial and once more went down to treetop level and we hedgehopped all the way across France to the Cherbourg Peninsula, which the Germans were holding. All guns were pointed upwards, so there wasn't much danger of getting shot down, but there was danger of being shot from below by some gunner or machine gunner. As we approached Cherbourg there was a huge cloud of dust that obscured the ground and we thought the Germans were attacking the Americans. I was going to report it so I broke radio silence, but before I got a couple of letters out, I was told: 'Get off the air you bloody fool.' So I did. We got away from Cherbourg, climbed up to a normal height and as we approached Ludford, we reported in and I got a message telling us to divert. I told the skipper and he said, 'Where to?'

'I looked at my code book and it said, 'Divert to Leicester East.'

'Right alongside the code for Leicester East was a code for Castle Donington; one letter difference. I told the skipper. We had a crew conference and the pilot said, 'What will happen if you tell us the wrong place?'

'Oh, they'll give me a wireless test; it'll be alright.'

'We landed at Castle Donington.

'This had been the longest daylight raid of the war I believe. The moment we landed there was a tractor to tow us to dispersal far away from prying eyes and huge tarpaulins were put over the entire aircraft, so that no one could see it. People had seen us coming in, but we strutted

---

11. 306 Lancasters of 1, 3 and 8 Groups attacked three oil-storage depots on the Gironde, at Bordeaux, Blaye and Pauillac – with excellent results. One Lancaster FTR from the raid on Pauillac. Thirty Mosquitoes of 100 Group escorted these forces without loss. *The Bomber Command War Diaries: An operational reference book 1939–1945* by Martin Middlebrook and Chris Everitt (Midland 1985, 1990, 1995).

around that place as if we were God's gift to the RAF. Operational types landed at an OTU. The instructors at Castle Donington had been on 101 Squadron, so we had a good laugh about it, but to us it was a special occasion.

'The day after, we were told that we could report back to Ludford Magna. The aircraft was uncovered, they drove us round to it and we started up, proceeded round the perimeter track and as we approached the take off point, we realised that practically the whole station was out, lining both sides of the runway to see this strange aircraft with a big mast. We took off and they were waving and cheering to us. We were so very grateful for the reception we'd received. The skipper said, 'I'm gonna go round and do 'em an overshoot and waddle the wings.' So we went back around and we came along the line of the runway. There were crowds of people all the way along the runway, both sides. He waggled his wings for them and we waved to them and they waved back to us. We returned to Ludford and reported to the intelligence officer with the result of our bombing and it appears that we'd done a very good job. We'd flown at such a low level and fifty miles from the French coast too, we were actually flying below the radar screen all the way down there before we bombed. Of course, they knew we were there after we'd bombed, but that's another tale.

'On 7 August we were scheduled to go to Fontenay-Le-Marmion. We took off and we were almost at the target when we were recalled and it wasn't counted as an operation. On 9 August we were told we were going to bomb Fôret de Nièvre. There had been sightings of German tanks going into this forest and no tracks coming out, so obviously, the German tanks were in the forest, probably waiting in ambush. This was a night raid. We took off in LL756 again and made our way to this forest. On arrival, our orders were to bomb to the right of the red indicator. Most of the bombs were incendiaries. When we got back we learned from intelligence that it'd been a very good raid and that the forest was completely destroyed. No German tanks had been seen to come out of the forest.

'On 12/13 August we were sent to Brunswick, but at briefing we were told that we would navigate by dead reckoning, so as to make sure we got to the target. We'd 13,000lbs of bombs on board. We set off for Brunswick with all the usual palaver on the way there, but immediately after turning for home, the navigator said: "Jack', have you heard any broadcasts about high winds?'

'Well I've got indication that there's a very high wind; a cross wind and a headwind.'

'He estimated that it was about 200 mph, which meant that instead of going forwards, we were going sideways. We were going slightly forward and eventually, we were coned in searchlights over Bremen, miles from our usual course. We were in the searchlight cones and the ack-ack was firing at us and we were in these searchlights for eight minutes. All the while, the rear gunner reported that there were two fighters shadowing us. Of course, they wouldn't attempt to attack us while the flak was coming up, but after eight minutes, the skipper decided that he would dive. He said, 'Hold tight lads, I'm going into a dive to get out of these lights.'

'He did tell me that when he dived, the needle went off the clock, which I think was about 400 mph, so he didn't know how fast we were diving, but we got out of the lights and when we eventually folded up. Others on the course across hadn't. Suddenly, there was an almighty bump. We didn't think we'd been hit, but it felt as if we had. We carried on and made for home. The ground crew said, 'There seems to be a lot of little holes, but we can't see anything else.' We thought, 'That's not bad, none of us was hit.' The wireless maintenance man said, "Jack', I want you to come with me a minute. You reported a bump. Well, there's a huge piece of shrapnel that's entered the fuselage, just forward of the rear turret and gone along the bomb bay, making a bit o' wreckage and through the top o' the bulkhead and settled into your seat.' I didn't know this.

'Would yer like it as a souvenir?'

'No thanks; you have it if yer want.' So he kept it. In addition to the shrapnel, someone pointed out to us that there were bits of twigs and leaves in the engines, which suggested that we'd had a further stroke of luck, because we must have pulled out of a dive very close to the top of some trees.

'Tuesday the 15th of August was a daylight operation [1,004 aircraft attacking night-fighter airfields in Holland and Belgium in preparation for a renewed night offensive against Germany]. Our bomb load consisted of very high explosives of various types to a total of 13,000lbs; and we had a Spitfire escort. We climbed to about 16–17,000 feet, met the fighters over the Channel and proceeded to Holland. There was a night fighter aerodrome with lots of night fighters about. I'm afraid we spoilt their lunch because we arrived overhead at about 12 o'clock. The Master Bomber marked the target and we were ordered to bomb a certain marker that was on the intersection of the three runways. This we did from 17,000 feet and with devastating effect. All our bombs appeared to hit the target area. We were so pleased about this, but then we had to circle around because our job was to do a line overlap to prove what the bombing had done. We had to wait until the last aircraft had bombed and then we did our line overlap. We were escorted once more by the Spitfires, so we felt quite safe. The scene below seemed totally devastated; it didn't seem as if there was an aircraft anywhere that hadn't been touched.[12]

'Wednesday 16 August was a very dangerous night trip to bomb the port of Stettin. There was an indication that wherever possible, paddle bladed props, which gave a distinct advantage when gaining height, were to be fitted, but it was not to be. We had to go and do the job, so off we went to do it. We set out across the North Sea towards Denmark, across a part of Norway, up to Sweden, at a very good height – above 20,000 feet. Our route was to take us down Sweden and across the Baltic to bomb Stettin. Swedish fighters came up alongside to escort us and make sure we didn't go any further and probably to help keep the German fighters from us. So, we crossed the Baltic Sea and there were already fires burning when we got there, but we dropped our bombs and there seemed utter devastation; there were fires all over the place and as we turned and headed back towards Sweden, we could see them for miles and miles.

'As we reached Sweden again, the fighters escorted us once more to our turning point and we turned to come back along the route we'd travelled. Although there were odd fighters about, no-one attacked us whilst we were over Sweden. We came home across the North Sea and relied on 'Jock' to get us back to Britain at least. We knew it was touch and go whether we'd arrive at Ludford or not, or if we'd run out of fuel but he got us back to Ludford. When we landed, we had five minutes' fuel left. This was to be the last trip before paddle bladed props were to be fitted. We were on standby whilst this was being done – to fly in other aircraft if needed, as was the case. During the Normandy campaign, we

---

12. A force of 110 Lancasters and four Mosquitoes heavily cratered Volkel and put both runways out of action.

had to be on standby. We never stood down as a squadron all throughout this period.

'We go forward now to Tuesday night, 29 August and were scheduled to return to Stettin. This time it was a little different because approximately one hour before we were due to set off and head for Poland, there was another large raid that was going to Königsberg in East Prussia. These were to take the same route across the North Sea and turn round by the Baltic so we expected that when we got into that area, the fighters would already be up. They were as we approached Demark but none attacked us. Our Lancaster had paddle blade props fitted and it was lovely to be able to get up to 29,000 feet on occasion and avoid a lot of the damage. There were huge bangs and bumps as ice fell from the propellers and each time this happened and the pilot had to settle for a lower position until the ice cleared and then up again. It was a see-saw trip, but we managed it all right. We made our way across Sweden and came down and bombed Stettin, but as we set out again for Sweden and the North Sea, once again the fighters were in evidence but none attacked us. When we got back to Ludford, they burnt the fuel out to a 'T', but we were comfortable. We got back, as did most of the others but Pilot Officer Piprell was one who did not return from this trip.[13]

On 31 August we were off once again, this time to Ste-Riquier, in LL756 once more, our own SR-Z. This target was a dumpsite for the latest V2 rockets which had been reported by the French people as to where it was. We arrived back at 1717 and we were quite pleased about the good job we'd made because the V2s were fired up into the atmosphere and they weren't particular as to whereabouts they exploded, just so long as they came down on London.

'Off once again, on 3 September, to Gilze Rijen aerodrome in Holland. We went off at 1553 and down again at 2013. I'm pleased to say that once more, we made a good attack and we got back safely. On 5 September, we went to Le Havre to bomb the German headquarters. We made a good job of that, total flying time was only three hours and 25 minutes. We

---

13. Pilot Officer Gordon Leslie Piprell RCAF who was 22 years old and five of his crew – average age 20 – were killed when Lancaster ME592 SR-D was shot down on 29/30 August. The sole survivor evaded. 23 Lancasters were lost on Stettin. At Königsberg 15 Lancasters FTR. *The Bomber Command War Diaries: An operational reference book 1939–1945* by Martin Middlebrook and Chris Everitt (Midland 1985, 1990, 1995).

went to Le Havre again, in LL756 SR-Q, on 6 September. This time we took off at 1748 and came back down at 2050. We had a full complement of bombs on board and as we approached the Channel, we were told by the Master Bomber to return, as enough damage had been done the previous day. Our skipper said, 'Right lads, I don't like this idea. We've got 13,000lbs of bombs on board, but we'll see if we can be first back because we don't want the whole squadron on this job.' He anticipated that there would be some trouble when we got back. After three hours and five minutes, we were home. We'd raced to get back and we were first back. But as we got back to Ludford we could see the windsock blowing, so the pilot asked for permission to land. They said, 'Yer can land on runway two-zero.' The skipper queried it because they were landing us downwind. A voice said: 'This is Group Captain King; you will land as ordered.' King, at that time, was in his bungalow in the village, but he was connected up and he interceded with flying control orders. The skipper said, 'Now lads, I don't think we'll be able to manage this, so, as soon as we touchdown, get into crash positions because I know we shan't be able to stop. Flying into the wind, we may have had a chance but flying downwind, no chance' so the order from the skipper to the engineer was 'wheels up.'

'We belly landed, bumping and banging along, sparks flying outside; we didn't know what was happening. It was impossible for us to stop and we were soon off the end of the runway taking the FIDO pipes with us, rising over the crest of the hill and down towards the 'Black Horse' in the village where we came to rest. I'd already anticipated that we had to get out quickly, so I was half way along the fuselage when Don dropped out of his turret in front of me and we went out of the door and passed Colin along the way; he was still in his seat. Out of the door and then we set off running. Don said, 'Ey up 'Jack'.' Ginger, the rear gunner had turned his gun turret sideways, so that he could roll forward over the guns and out but Don said that 'Ginger' was stuck. We both turned back. I got there first and I gave the quick-release on his armour a bang and down fell Ginger between the guns onto the floor. We knew other aircraft would be landing, but all our thoughts were of the bombs on board we could see that they were glowing red hot beneath the aircraft and we ran across the runway to flying control. *WE* were the first to do the four minute mile, not the famous doctor – with our harnesses on! We were already drinking cups of coffee when the rest of the crew arrived on a crash tender. The aircraft was a complete write-off.[14]

'On the night of 12/13 September we were detailed to go to Stuttgart in Lancaster III NF954, a brand new 'W'. Once again, we knew this would be a long trip. We set off at 1855 towards southern Germany and came down at 0222 – a flying time of seven hours and 27 minutes. It was a very good attack; we hit the works and industrial areas we'd been sent to do and so now, we were well on the way to finishing our tour.

'The next trip was in NF954 to the Ruhr. It was a short trip – three hours and 51 minutes – because it was straight there and straight back. The bombing was very concentrated and we did a very good job. On 17 September in the same aircraft, we bombed Westkapelle [one of eight locations to the east and west of the harbour town of Flushing], an island where big guns were making the supply lines very difficult up the coast. Half the bombs undershot the target but some knocked the guns out and the others made a break in the sea wall, which, when further raids were made, caused considerable flooding. On 23 September we went to Neuss. This was likened to a place the size of Rotherham by the briefing officer, when he gave his information. It was very concentrated bombing and I'd be surprised if anyone was alive after this, because it was a large force that went there and we all dropped our bombs on the target and there were fires and all sorts going on down below.

'On the next operation, on 26 September, we were in support of the people of Dover because the guns that had been firing across the Channel were at Cap Gris Nez. We were up at 1114 and down at 1349. It was intended that once and for all, the guns at Cap Gris Nez should be silenced. Once again, this was an NF954. It was a wonderful attack and what pleased us most was that on that occasion, the gunners were allowed to fire at anything down below, if they saw anything. I don't know whether our gunners did or not, but it was a concentrated attack. I'm pleased to say that in the *Daily Express* the next morning, there were photographs on the front page of guns with their barrels broken, all down on the floor. This was great because these went back and forwards on a

---

14. Lancaster III ND983 SR-B flown by Flight Lieutenant A. Massheder also crashed at Ludford on return. There were no injuries. W. R. Chorley.

railing; they sheltered under the cliffs, whatever, we managed to cop 'em that day.

'On 5 October we flew in a brand new Lancaster, NG129 SR-Z; a replacement for the one that we'd crashed on 6 September. I didn't go on this one, but the rest of my crew did and I understand it was a very good mission. Three days later I walked across to the wireless section to find out if I could get on a spare trip to catch up with my crew. As I walked in, Flying Officer Roland Holmes, who was one of the pilots on the squadron, said, 'Oh, hello 'Jack', what yer doing here? I'm looking for a wireless operator for our last trip.'

'Well, I've come to volunteer for a spare trip, so I can catch my own crew. They went to Saarbrücken and I want to do a trip to catch them up.'

'Oh, we'll be very pleased to take you with us.'

'And that was it. The target this time was Bochum, a steel town similar to Sheffield. We were off at 1732 and down at 2228. We bombed the marker flares and we were supposed to be bombing the oil tanks, but I think we hit most of the steelworks as well but, there were numerous fires among the oil tanks. I was a little uneasy on this trip because this crew wasn't as disciplined as my own crew who only spoke when they needed to. This lot were shouting and talking to one another all the time and we almost got shot down because I was giving direction of a fighter coming in to attack us and the two gunners just weren't interested. The pilot told them to shut up and listen to what I'd got to say. They turned their guns towards the fighter attack and he didn't attack but I was worried about getting back, especially with this crew being on their last trip.[15]

'Before I'd chance to once again fly with my own crew, on 14 October I was detailed to fly with Flying Officer Simpson, a Canadian, on a 'daylight' on Duisburg [by 1,013 aircraft – 519 Lancasters, 474 Halifaxes and twenty Mosquitoes in Operation 'Hurricane']. We took off at 0622 and landed back at 1110. Simpson reported that he'd made a good attack. Well, this was not the case because as we approached Duisburg, we were given the codeword 'Freelance' by the Master Bomber, which meant: 'Don't bomb Duisburg; look around for another target to bomb.' We did this and we saw a circular looking town with streets running off to the

---

15. Every reunion Holmes would come and shake my hand and tell people, 'This lad helped me to get through my tour.' I was so proud of that.

centre but every other road was circular. I know this because while they were talking about it, I got up and had a look. We bombed a building in the centre.[16]

'The next trip was with my own crew on 15 October and we were in NG129, our new aircraft. The rest of the crew had flown in it to Saarbrücken. This time it was to be the port at Wilhelmshaven – off at 1740 and back at 2146. This was a very good attack for us.[17] Later, I borrowed a pilot's bicycle and although there were no white lines or other indication, no cat's eyes on the road, cycled the twelve miles to my wife's lodgings in Louth. I could just discern the difference between the tarmac and the grass at the side of the road and so I went into almost every gateway on the way, but I got there.

'On 19 October in NG129 once more, off to Stuttgart again, at 1711 and down at 0000. We carried one 4,000 pounder and a load of high explosives. The usual good attack without any interference from the Germans and it was really a pleasure to get back. The skipper said he'd had a 'marmalade' from the master bomber, which meant, 'turn back and don't bomb', but I told the skipper that there was no marmalade that day, because we'd already had 'marmalade' on a previous trip and the Master Bomber never used a code word twice. This was the Germans trying to turn us back, but after an argument with the skipper saying 'who's pilot of this aircraft?' I insisted that if we turned back, I'd be reporting the skipper and of course that would have meant a court martial, so he decided to listen to me.

'The next trip, on 23 October, which we expected to be our last one, was Essen, kicking off at 1634 and back down at 2249, but the thing went wrong. At the coast the starboard outer started playing up and had to be feathered. We had a crew conference and we decided that we'd carry on, but then one of the other engines started playing up. We approached Essen and the bomb doors were opened but we couldn't drop the bombs, probably due to this starboard outer. We circled the target and abandoned the mission and turned for home. The bombs were primed. We tried pumping the lever by the side of the wireless operation position, which

---

16. 957 bombers dropped 3,574 tons of HE and 820 tons of incendiaries on Duisburg. Fourteen aircraft – 13 Lancasters and a Halifax – were lost.
17. 506 aircraft – 257 Lancasters and eight Mosquitoes from all Groups except 5 Group claimed 'severe damage' to the business and residential areas.

when pumped backwards and forwards would enable the bombs to be dropped, but in our case, it didn't happen. We were in a bit of a stew now because the starboard engine started kicking up and had to be feathered again. As we neared base we were told to go to Carnaby, a grass airfield south of Bridlington with a five mile long crash land strip.[18] The navigator informed the pilot that we were just over the cliffs and that they were fifty feet high, so the pilot in his wisdom decided, 'Right, then we go out to sea and come back in with a trailing aerial out. We shall then be flying at fifty feet, so it's very dicey', but we made our way in and he sez, 'I'm not going to try and land it: I'm going to fly it on.' And he did just that; it was a really brilliant piece of piloting, more so because as we were approaching this crash landing strip, we heard a broadcast which said, 'Lancaster coming in to land; may explode. Clear the airfield.' Well that didn't sound very promising. The skipper came to the fore and we all did as we were told.

'I said, 'Are we down yet Skipper?'

'Yes, I'm going to just taxi off the end of the runway and cut the engines and let the ground staff look after us.'

'Well, that's brilliant, after the way I'd slated him about not being able to land, about thumping up and down and when we needed him, he just turned up trumps. Next day they got the bombs off and got the aircraft fit to fly back to Ludford. We were pleased about that because we knew we would have to fly another trip in place of this one. It should have been our last. We'd been over the Ruhr Valley for nothing.

'On 28 October we were on a 'daylight' again, in NG129. The target was to be a bridge across the river at Cologne. Our orders were that whatever happened, we must bomb the bridge by proceeding down river or up river; I don't know which. We must not let any bombs fall on the cathedral, but we must put that bridge out of action. We were very high when we dropped our bombs. The rear gunner reported, 'OK, it's a direct hit, we've hit the bridge.' Later our photographs showed that we had indeed hit the bridge, but on our target photograph there was another aircraft below us. It was the famous 'S-Sugar'. Whenever I met Jerry Murphy after that, he was slated for trying to kill us! 'Sugar' lasted until the middle of March before it got shot down. Well, we did not bomb the cathedral. When we

---

18. An emergency airfield with a 9,000 feet runway. It was also a FIDO airfield.

got back we had a committee waiting for us as we taxied to dispersal.[19] The squadron commander was there with a huge yellow flag depicting the name, '*Harris' Rebels*' a skull and crossbones. We learned that this flag had been made of ties, cut off from the officers in the officers' mess by the WAAF officers and then stitched together. We had our photograph taken with our ground crew after the wing commander said, 'Now then Morley, you're the biggest rebel of the lot; get hold of this and get on the front row,' which I did. At long last we'd finished our operations. The wing commander had deposited some money in the sergeants' mess – 'enough to last yer all night, so get yer friends and enjoy yer first night without worrying about going on operations'.

'The skipper came round the day after and said, 'The three officers have been awarded the DFC' and there's one DFM to be awarded among the sergeants.' He came to me and said, 'Would you like it?'

'I wouldn't. I've already had one DFM taken off me for being a naughty boy.' I was recommended once before but I went to watch a football match and they found out I was away so, that was that.'

'All right then, if you don't want it, I'll go and see the others.'

'He went round all the five sergeants and every one of them refused it. I was so pleased and yet, I was so sorry for little 'Jock' the flight engineer; if anyone should have got a medal, it should have been him. I missed one trip with the crew, but he'd been with them all the time. In February at Ludford Dai Jones, the bombing leader at the time was sporting his DFC. He told me that John Arthur, the navigator was instructing.

At the end of May, the war in Europe was over. Then, when VJ-Day was announced, I was in transit somewhere, but I got my own back on Flight Sergeant Frazer because I had to report to Blackpool for a couple of days, until they got a permanent warrant officer and then I was to move on to Bruntingthorpe, near Leicester.

That first morning Frazer came in and said, 'Hello.'

'Just stand to attention when you talk to me.' He stood to attention.

'Take that white flash out of your cap.'

'Well, I always wear this.'

---

19. 'Sugar' was shot down by a Me 262 jet fighter on 23 March 1945 in the attack on railway bridges at Bremen and Bad Oeynhausen. Three of the crew including the pilot, Flying Officer Ralph Robert Little RCAF, were killed.

'That's for air crew under training and you're not air crew under training.' So he had to take it out.

'I've got a message for you to take down to Squire's Gate.'

'Shall I get transport?'

'No, go down on your bicycle.'

'While he was going to Squire's Gate, I got my new orders. A new warrant officer had arrived and took over from me and I moved on towards Bruntingthorpe where I was put in charge of German and Italian prisoners of war working on farms. The Italians would grumble, but I made sure the Germans got better treatment than they did.

'I was finally de-mobbed on 28 September 1946.[20]

---

20. Adapted from a story submitted to the People's War site by Bill Ross of the 'Action Desk – Sheffield' Team.

Chapter 4

Lancaster Love Story

Dear Skipper,

I take the liberty as referring to you as above, as that is how Johnnie used to talk of you.

My wife and I thank you very much indeed for your kind references to our boy and who should know better than you what kind of boy he was. He was heart and soul in his job; in fact it was his life. We get a little consolation from the fact that he did what he set his heart on doing and if he 'went out' doing it, he did it in the full knowledge of the risks entailed. It is a terrible blow to us as you may well imagine. He was a grand boy who was content in his home life and had never, at any time, caused us a moment's trouble. He would, I am sure, have gone far in civil life. We miss him sadly, especially Mary, our daughter. They were never without each other. Both she and my wife are taking it wonderfully, but I tell you what I have told others. I cannot take it. I served four years in France during the last war; all front line service and had my pals killed at my side and should be hard but it is altogether different when it comes to your own. Should you ever get a chance to call on us, we should be very grateful to have a chat with you…

Letter dated 20th June 1944 to 'Steve' Stevens on 57 Squadron from Mr. Smith in Woolwich, London, whose 20-year-old son Sergeant John George Smith the rear gunner on 20-year-old Pilot Officer Joseph Howe's crew on 630 Squadron was killed on the Berlin raid on the night of 23/24 November 1943. There were no survivors and their names are commemorated on the Runnymede Memorial to the Missing. 630 Squadron was formed at East Kirkby on 15 November 1943 with eleven Lancasters from 'B' Flight of 57 Squadron (which included five Lancasters that had been repaired two or three times following damage) that had arrived from Scampton in August 1943.

Sidney George Stevens, who was always known as 'Steve' was born on 14 December 1921 in Torrington in rural Devon at a time when many people still travelled by horse and cart. 'At school, if we heard an aeroplane we'd all run out of the classrooms to try and see it' recalls 'Steve', who did not imagine he would ever fly in an aeroplane, let alone take charge of one. His father was an engine driver on the Southern Railway and when 'Steve' was 12 the family moved to London. With money short, 'Steve' left school early for a job in the housing evaluation department at Croydon. But with war looming the teenager fitted his day job around voluntary night shifts in an Air Raid Precaution control centre training to co-ordinate the clear-up after bombing raids. When the Blitz began he found himself in charge of one of the centres. He received no pay but was given a shilling (5p) a night for tea money. He would never forget the air raids or the wrecked houses. During the day he had to requisition houses for those whose dwellings had been destroyed. One night, he was starting his 8-hour voluntary shift when a message was received that his family home had been hit by a bomb that had exploded in the street. He stayed at his post until relieved at 6 am and then cycled home exhausted over fire hoses, bricks and rubble. There was just a gaping hole where his home had been.

'It was fortunate that my parents took a stout kitchen table when they moved to London. They had dived under it and were dug out alive when a German bomb had demolished the house. The kind lady who lived next door was plastered to the wall of her house like some hideous gelatinous graffiti and the bruised face of my father, the white bomb dust in my mother's hair and the broken leg of an uncle who had not managed to squeeze all of his body under that table etched themselves on my mind forever. Everything for which I had worked and saved was pulverized except for a torn shirt, a pair of Boy Scout shorts and a school prize dated 1935. These were my only possessions apart from the clothes I was wearing and the bicycle I had ridden to work. I stood in the back garden and shouted, 'You bastards. I'll get my own back on you'. My spirit demanded revenge and I was determined to repay that foul deed.' He tried to volunteer for pilot training in the RAF. 'They had thousands of people wanting to be pilots. I was turned down because I was too young. I didn't become 18 until right at the end of 1939 because I was born in December. I put in my papers the following April and was sent for a medical that summer but was not called up until about July 1941. I was amazed to get through.'

Months of intensive training and testing followed. He learnt about everything from the mathematics of navigation to how to build latrines in the desert and finally began flying. To his surprise, in 1942, he was then sent to America where he learnt to fly training aircraft like the PT Stearman biplane at No.2 BFTS War Eagle Field, Polaris Flight Academy; a small British flying training school at Lancaster in California. In 1938 the Army Air Corps had recognised that it would have monumental problems in developing a tremendously expanded Air Corps. By 1939 General 'Hap' Arnold, its chief of staff, realised that the USA had to plan for the possibility of involvement in the European war and a great responsibility to help its ally, Great Britain. Arnold's plan was that the primary training phase of flight instruction be placed with civilian operated schools where all services and facilities, except the aircraft, were furnished by the operator but with AAC control of the methods and manner of the instruction. In the spring of 1939 Arnold called eight successful civilian pilot training school owner-operators affiliated with the Aeronautical Training Society to Washington DC. All agreed with Arnold's suggestion that they become contractors with the Army to provide primary pilot training for 12,000 pilots per month. By July 1939 nine civilian schools were giving primary phase flying training to AAC Aviation Cadets. By August 1940 nine more schools were in operation (by the end of 1940 Arnold's ambitious expansion programme would be training more than 30,000 pilots a year). In 1941 seven British Flying Training Schools were established. All were civilian operated and run with civilian American flying and ground instructors and a small RAF supervisory staff. More than 7,000 pilots were produced for the RAF by the BFTS by the war's end.[1]

While at Lancaster the cadets were given leave to enjoy the Californian sunshine. Whilst off duty, 'Steve' Stevens and Len Gray his friend and fellow cadet explored Hollywood and Los Angeles. On 27 April they were admiring the Spanish mission at Santa Barbara when Alistair Cooke,

---

1. When the Arnold Scheme finished in February 1943 with Class 43-D, the thirteen classes had produced 4,500 RAF pilots from an intake of 7,500 RAF entrants.

the legendary radio broadcaster, introduced himself.[2] 'He took us for a ride along the coast and then to view some superb mountain scenery. We descended to the sea where he bought us lunch.' Len Gray subsequently wrote: 'Our stay of six months-plus at two British flying training schools at Lancaster was, I am certain, the best part of our wartime experiences for the ten courses (approximately 500 cadets) who learned to fly there.'

Both men soloed and returned to England in early 1943. 'Steve' Stevens, now a Pilot Officer, was posted to a OCU where he crewed-up and flew clapped out Wellington Ic's whose cowlings glowed red hot on night exercises. 'Flight Sergeant E. J. Howes was the navigator. My bomb aimer was Sergeant Lamblin; Sergeant K. D. Brown was the wireless operator and Sergeant James George Louis Martin, a Frenchman was the mid-upper gunner. Rear gunner was Sergeant Townsend.'

At 1654 HCU Wigsley 'Steve' Stevens' crew converted to the Lancaster and a seventh member of the crew, the flight engineer, Sergeant Eric Blanchard, joined them. 'From the 13th to the 23rd of May I flew by day and night continuing to gain experience by practice bombing, gunnery, navigation and instrument flying. The magnetic compass was 'swung'. The bombsight levelled and the gun turrets tested. The front gun turret had a faulty rotating servo joint which would spray the windscreen with light oil and on two occasions I had to land the aircraft at night peering through the small clear vision panel. This was a useful exercise as there were later occasions when the windscreen became so thickly coated with ice that forward vision was negligible.

Returning from a training flight at Wigsley 'Steve' Stevens was astonished to hear a woman's beautifully clear, steady voice. He was expecting to be given landing instructions by one of the men he knew in the control tower. Instead a woman's voice came over the airwaves. 'It was the first time I had heard a girl's voice from the tower, so I thought

---

2. Alfred Alistair Cooke KBE (20 November 1908–30 March 2004) was a British/American journalist, television personality and broadcaster. In 1937, Cooke moved to the United States, starting what was to become a permanent emigration. He became a US citizen and swore the Oath of Allegiance on 1st December 1941, six days before Pearl Harbor was attacked. During this time, as well, Cooke undertook a journey through the whole United States, recording the lifestyle of ordinary Americans during the war and their reactions to it. The manuscript was published as *The American Home Front: 1941–1942* in the United States (and as *Alistair Cooke's American Journey: Life on the Home Front in the Second World War* in the UK) in 2006.

I would go up and have a look! There was this glamorous girl sitting in the middle of a crowd of people, the centre of attention, so I crept away.'

Corporal Maureen (christened Maud Ellen) Miller was a 23-year-old WAAF radio telephone operator, born and brought up in Norwich, the daughter of a Regimental Sergeant Major in the Royal Marines and the youngest of seven surviving children. Her first job on leaving school at 14 was a copy-holder assisting manuscript readers at Jarrolds – normally a post for 16 year-olds with the School Certificate but she copied well for two years. A number of jobs followed until she became head cashier and book keeper at the world famous Hippodrome theatre taking advance bookings. When she turned sixteen, Maureen tried to enlist in the WRENs but was deferred because of a very minor toe defect before she was accepted for the Women's Auxiliary Air Force (WAAF) in 1941.[3] Maureen thought that she would be doing clerical work but in the interview they kept her talking for a long time and then said they had something different for her and she began R/T training at RAF Waddington. 'Pip' Beck, a fellow trainee, had an adjoining bed in the same hut as Maureen and took to her immediately. "Maureen was tall, slender and attractive, with thick, honey-coloured hair and green eyes framed with long, black lashes. I guessed her to be a year or two older than me, with a pleasant manner and an air of gentle, though positive, self-possession."[4]

Maureen meanwhile, had become a fully fledged R/T operator. 'As the pilots came back it was my job to land them, or tell them to circle in a stack and bring them in one by one in turn, unless one had been shot up and needed to land straight away. They loved to hear my voice and I loved to hear theirs. The first time I went into a control tower I was petrified. I had never been near an aircraft before. Then, my first night on duty, a plane took off and went up in flames. I went racing down the stairs, I just wanted to get out, but the man in charge pointed to my seat and said, 'Miller! Your place is there.' So I came back and sat down. It was a job I loved. I had never expected to leave Norfolk and there I was

---

3. Had Maureen stayed at the Hippodrome her life might have been abruptly ended. On the night of Wednesday 29/30 April 1942, forty Luftwaffe aircraft carried out a devastating raid on Norwich and the Hippodrome took a direct hit.
4. *A WAAF In Bomber Command* by Pip Beck (Goodall Publications Ltd 1989).

landing aircraft. Mayday calls could be very exciting. Crews would get lost or disorientated and we had to talk them safely down.

In the first week of May 1943 'Steve' Stevens' and his crew were posted to 57 Squadron at Scampton. Their skipper began his tour as a '2nd dickey' on a trip with 23-year-old Pilot Officer Jan Bernard Marinus Haye who had completed 22 bombing operations. Born at Pekalongan in Java of Dutch parents who divorced in 1929, Haye attended school in Hilversum and The Hague. After his mother remarried in 1938 and went to live in England he visited her on holidays and from 1939 he went to schools there. 'Dutchy' or 'Bob', as he was known, considered himself more English than Dutch. The night that Stevens was due to fly with the Flying Dutchman the raid was cancelled. Instinctively he thought that Haye would not be on the squadron for much longer.

On the night of 12/13 May, Stevens flew his '2nd dickey' trip with Squadron Leader Ronald Edward Sidney Smith DFC, on the operation to Duisburg, when 572 bombers (238 Lancasters, 142 Halifaxes and 112 Wellingtons, 70 Stirlings and ten Mosquitoes), took part. 'Squadron Leader Smith also took my flight engineer and my bomb aimer' recalls Stevens. 'Long before we arrived at the target the Ruhr 88 mm heavy batteies put up a huge barrage, blasting the air with heavy explosions which rocked the aircraft while lighter flak streamed up like flaming onions. The whole scene was lit up by searchlights which turned night into day. On our return we found that 34 of our aircraft had been shot down, including two from our squadron.[5] Several other aircraft were badly damaged [and three crashed in England]. With another 29 trips ahead, I thought that my survival chances were small.'[6] 'Dutchy' Haye in fact failed to return from his next trip, his 23rd.

On the night of 13/14 May 442 aircraft were detailed to bomb Bochum a highly industrialized and an important transport centre for the entire Ruhr and another 156 Lancasters and twelve Halifaxes set off on a long

---

5. ED329, which was shot down by Major Walter Ehle of Stab II./NJG 1 flying a Bf 110G-4 from St Trond airfield, Belgium, crashed at Maasniel behind the brickworks factory, Limburg. Only the bomb aimer survived on the seven man crew. Flight Sergeant G. B. Leach RCAF and four of his crew on ED778 survived after it was shot down by Hauptmann Wilhelm Dormann of 9./NJG 1 flying a Bf 110G-4 from Twente airfield.
6. Squadron Leader Ronald Edward Sidney Smith DSO DFC was killed on 9 September 1944 aged 28, when both engines on Mosquito PR.XVI NS565 on the strength of 34 Wing Support Unit cut on takeoff from Northolt.

eight hour round trip to bomb the Škoda armaments factory at Plzeň (Pilsen) in Czecheslovakia about 56 miles west of Prague. 'Dutchy' took 'W-Willy' off from Scampton at 1137 hours to attack Pilsen, carrying one 4,000lb HC bomb and five 1,000lb bombs. At 2322 hours near Albergen n the Dutch province of Overijssel his Lancaster was the first of three Pilsen-bound bombers claimed shot down over Holland by 25-year-old Hauptmann Herbert Heinrich Otto Lütje, Staffelkapitän, 8./NJG1 flying a Bf 110G-4 from Twente airfield. Born on 30 January 1918 in Abbesbüttel, the son of a farmer, after graduation from school, Lütje joined the Luftwaffe on 1 November 1937. He claimed his first aerial victory on the night of 6/7 September 1942 and on the night of 12/13 March 1943 he had achieved his 20th aerial victory. Lütje's total for the night of 13/14 May was six British and Canadian bombers which took his score to 29.

Haye's two air gunners were killed. The navigator, flight engineer, bomb aimer and the Canadian wireless operator were taken prisoner. 'Dutchy' successfully evaded capture in his native land and reached Amsterdam on foot and by bicycle where he received help from the Dutch Underground. On 30 May he assumed the identity of and used the papers of Rudolf 'Rudy' Franz Burgwal, born in Surabaya in the Dutch East Indies on 27 September 1917, who had escaped to England in 1941 to join the RAF.[7] 'Dutchy' Haye hid in the Burgwal family house in The Hague for the next eight weeks. During that time he met and fell in love with 22-year-old Elly de Jongh, one of the Underground workers a few doors away. Finally, on 26 July Haye reluctantly said goodbye to Elly and sailed to England on a barge with other evaders. 'Dutchy' eventually returned to operations flying on 83 (Path Finder) Squadron until the end of the war. After VE Day he returned to Holland to find Elly de Jong. She and other members of the Underground had been betrayed and arrested and sent to Ravensbrück concentration camp under sentence of death. Elly was transferred to Mauthausen concentration camp in upper Austria from where she was liberated before the sentence could be carried out. On

---

7. Flying Officer Rudy Burgwal was killed age 26 by enemy flak over the Departement de la Mayenne, Pays de la Loire on 12 August 1944 while escorting a Lancaster raid in a Spitfire XIV of 322 (Dutch) Squadron.

8 November 1945 she married her RAF pilot and they lived happily for over fifty years.[8]

'Steve' Stevens did not expect to survive the war. 'On almost every trip people just didn't come back. Sometimes they would arrive at a base and move their things in to a room one day and then go out on a trip and never return. Sometimes it was damned hard getting back. You'd have holes in the aircraft. You were flying in total darkness. In the air I was so busy I didn't have time to think of what might happen. It was only afterwards. There was a high probability that I wouldn't live. And I didn't have too much to leave; only my bicycle for my younger brother.

'The German flak was superb. Highly mobile flak protected the Army from air attacks as the Wehrmacht conquered Europe and by 1943 the barrage of heavy flak which protected every major city was formidable. To approach any city, especially those producing armaments, meant that our soft skinned aircraft had to penetrate that barrage which often formed belts several miles deep, every time. Generally, we tried not to take violent action to avoid flak and did not worry too much unless we could hear it explode, see close bursts or smell it. The heavy 88 mm gun was a superb weapon for anti aircraft and anti tank use and German industry produced vast quantities. As so many of these guns were employed to deter the RAF the Russian Front was starved of the one weapon that might have stemmed the Russian tank advance.

'When approaching the target flak would explode and hurl the aircraft about while from below light flak would hose up menacingly. Surrounding the intense flak barrage would be hundreds of searchlights. Some of these were radar predicted. When an aircraft was caught it was speedily enveloped by a huge cone of searchlights and every available gun would fire at it. It would swerve, dive and climb like a moth in a flame and all too often there would be a blinding flash as it blew up its

---

8. See *Shot Down and on the Run; True Stories of RAF and Commonwealth aircrews of WWII* by Air Commodore Graham Pitchfork (The National Archives, 2003 & 2007) and *RAF Evaders: The Comprehensive Story of Thousands of Escapers and their Escape Lines, Western Europe, 1940–1945* by Oliver Clutton-Brock (Grub Street 2009). Two weeks after being credited with the six victories, Lütje was awarded the Ritterkreuz and he was appointed Kommandeur of IV./NJG6 in Rumania. He ended the war with 53 victories, 51 of these by night. Following the rearmament of the Federal Republic of Germany, he joined the German Air Force in July 1957. He retired with the rank of Oberst (colonel) and died on 18 January 1967 in Cologne-Wahn.

bombs and fuel making a massive explosion which few aircrews survived. Meanwhile German night fighters buzzed overhead their cannon at the ready to seek revenge on those trying to destroy their homeland. Several people asked me what aircraft I was flying. When I said 'Lancasters' the reply was always something like 'Good boy! Pay the beggars back! Give them hell!' In 1943 the air battles of the Ruhr, Hamburg and Berlin were fierce and at times, went heavily against us.'

'Steve' barely expected to survive each day – let alone dare to hope he might spend the rest of his life with the beautiful young woman whose voice guided so many to safety. It seemed that they were destined never to meet, but Maureen Miller had noticed him at Wigsley and in the second week of May, when by chance she was posted to Scampton, they met soon after while walking. A first date was arranged but 'Steve' failed to turn up. He had been detailed to go on another raid and had forgotten to ask her name! It was only when he heard her voice on his cockpit radio a second time, that he persuaded her to give him another chance. Walking near the airfield one evening it started to rain heavily and they took shelter in the porch of Welton church. 'Steve' proposed to the glamorous blue-eyed blonde with the beautiful voice he had first heard guiding him back to land. 'I think it went something like, 'If I'm alive at the end of the year, we'll get married!' Maureen said. 'I wasn't the marrying kind. I hadn't wanted to get married. I was the youngest of seven and had four older sisters and I used to say I didn't want to be like them and spend my life with one person! But I remember thinking, very early on, that he'd make someone a good husband!'

Each night Maureen relayed information from the aircrews to the senior commanders and talked Lancaster crews down at Scampton. "The work in the control tower was strictly controlled, everything being for the safety of the air crews. Pilots under instruction were in touch with the control tower all the time and my fellow workers and I, wearing earphones were instructed by a control officer. Planes taking off on raids did not contact the control tower – it was only on return that they would radio in. Having taken off, the plane's letter was put on a board, and next to the letter, the pilot's name and rank. 'Steve's aircraft was 'E-Easy'. On return, the time of landing was put on the board. If the plane did not return, it was left on the board and then rubbed off the next day. Many planes did not return. A log was kept of all transmissions and calls received. On return, the first plane to call in was told 'Pancake' and the station call sign ('Heron') and squadron call signs were used, but no names were ever used. The second to call in was told 'Aerodrome 1000' which meant he could circle at 1,000 ft. Subsequent arrivals were stacked at 500 foot intervals. If an aircraft radioed in that it was in trouble or badly damaged, then others were called collectively and asked to stand by. They were held in the air while the plane in difficulties was called in first. Sometimes there were calls from aircraft that were lost and the call sign for this was 'Darkie'. My job with these was to verify their identity and then guide them in to land. I found this a very interesting part of my work."

WAAFs also had the sad task of listening out until the last for those who did not make it home. At Scampton on the night of 16/17 May Maureen was the duty operator in the main control tower from midnight to 8 in the morning when she talked down some aircraft that had returned from a raid. "This particular night there seemed nothing out of the ordinary but as I left the end of my shift there was an air of elation and excitement everywhere on the camp but I did not know why. It was only later that I found out that I had talked down the Dam Busters, some of whom, including Guy Gibson, I knew slightly." Of the nineteen Lancasters on 617 Squadron which took off from Scampton on the Ruhr Dams raid, eight did not return and 53 of the 133 aircrew involved were killed; a casualty rate of almost 40 per cent.

'Steve' Stevens' next op, his first trip as a Lancaster captain, was on 23/24 May when, after a nine-day break in operations, the target was Dortmund in the Ruhr Valley area of Germany. Bomber Command detailed 829 aircraft to go to the region so sarcastically nicknamed 'Happy Valley' for what would be, except for the 1,000 bomber raids of 1942, the largest raid of the war. No less than 343 of the aircraft involved were Lancasters and the rest comprised 199 Halifaxes, 151 Wellingtons and 120 Stirlings and thirteen Mosquitoes. There was no element of surprise in their favour. As the bombs left the aircraft Stevens felt a moment of exhilaration. "Although two fighters made a pass at us, we returned safely. Two aircraft on my squadron were shot down.[9] After this, I had had my

---

9. Ground defences in the target area and night fighters on the return route put up a strong opposition and 38 aircraft, eighteen of them Halifaxes, failed to return. Eight Lancasters, six Stirlings and six Wellingtons were also lost.

fill of hatred. Quite frankly I had wanted to kill as many of the bastards as I could! Now I thought I had paid the debt I owed. We were trying to flatten German industry, or at least damage it. My crew and I enjoyed nine days leave. I wondered if we should survive to enjoy another."

In June 'Steve's crew flew three ops in quick succession – "quite a hard breaking-in for a new pilot who has just joined the squadron" – to Düsseldorf on the 11/12thth, Bochum on the 12th/13th and Oberhausen on the 14th/15th. The first two trips were on ED946 and the other on ED931, when they were attacked by a Ju 88 and a Bf 110.[10] On 12/13 June 14 Lancasters and ten Halifaxes from the 503 aircraft sent to bomb Bochum were destroyed by flak and fighters. The raid took place over a completely cloud-covered target but accurate 'Oboe' sky-marking enabled all 323 Lancasters and 167 Halifaxes to cause severe damage to the centre of the city and 130 acres were destroyed. On the night of 14/15 June on the outward flight to Oberhausen, Martin – Stevens' French mid-upper gunner – damaged a Messerschmitt 110 and on the return flight, a Junkers 88. 'It was a fairly classical attack' Stevens recalled. "The 110 came in and did all the things we hoped he might do in accordance with what we'd been told to expect. I took the necessary evasive action; the gunners fired back and although he did get some hits on our aircraft they managed to shoot him down and that is not an easy thing to do. We only had machine guns and the 110 had cannon and machine guns so the display of fireworks up front was pretty uncanny! When we were on the way back I thought we were all right but our starboard outer was misbehaving a bit so I thought we'd better feather it. Then the coolant ran out and the radiator temperature went up so we flew back on three engines." Martin was later awarded the DFM, his citation praising him as "a most reliable air gunner who invariably displayed keenness, determination and cool courage." Oberhausen was cloud-covered but once again the 'Oboe' Mosquito sky-markers were accurate. Seventeen of the 197 Lancasters despatched failed to return.

---

10. ED931 went MIA with Flight Sergeant Ernest Frank Allwright and crew on the raid on Hamburg on 29/30 July 1943. All seven crew died. ED946 was being piloted on a training flight by 21-year-old Sergeant John Basil Josling of 1660 HCU on 28 August 1943 when out of fuel, it crashed one mile from Swinderby on take-off killing all nine crew. Josling left a widow, Patricia Mary Josling of Lydiard Millicent, Wiltshire.

'Steve's' crew flew their sixth operation on the night of 16/17 June when the target was Cologne. Fourteen Lancasters were lost from a force of 202 Lancasters and ten Halifaxes of 1, 5 and 8 Groups that raided the city. On 21/22 June, before the moon period was over, the crew flew their seventh operation when Krefeld, just inside the German-Dutch frontier was the target for 705 aircraft (262 Lancasters, 209 Halifaxes, 117 Stirlings, 105 Wellingtons and 12 Mosquitoes). A total of 44 aircraft (17 Halifaxes, 9 Lancasters, 9 Wellingtons and 9 Stirlings) were lost; 38 of them to night fighters operating mainly over the southern provinces of the Netherlands. The Path Finder aircraft produced an almost perfect marking effort, ground markers being dropped by 'Oboe' Mosquitoes being well backed up by the Path Finder heavies. In 53 minutes 619 bombers dropped 2,306 tons of bombs on the markers, more than three quarters of them achieving bombing photographs within three miles of the centre of Krefeld – approximately 47 per cent of the built-up area – which was burnt out and over 5,500 houses were destroyed and 72,000 people lost their homes. Twenty thousand of these were billeted upon families in the suburbs, 30,000 moved in with relatives or friends and 20,000 were evacuated to other towns.[11] All told, Bomber Command lost 275 bombers shot down in June 1943.

'Steve's' crew flew two two trips to Cologne – on 28/29 June and 8/9 July for their 9th and 10th operations. On the 28th the met forecast was that there would be cumulonimbus clouds up to 20,000 ft over the target. 'The weather was expected to be unsettled and only Lancasters flew. The target was cloud covered and the sky marking was good. We saw a small cluster of red target indicators and bombed and then set course for home. It was then that a loud crash shook the aircraft and there was a smell of cordite and dust and a great deal of noise. We were clearly hit by something large as bits penetrated the aircraft. (A shell had gone through the bottom of the aircraft immediately under the mid-upper turret). I called the crew in turn. The gunners did not reply because the intercom had been cut. I asked Sergeant K. D. Brown the wireless operator to see if he could find out what had happened at the rear. He went back and I lost contact with him. After a while I asked Sergeant Lamblin the bomb

---

11. *The Bomber Command War Diaries: An Operational reference book 1939–1945* by Martin Middlebrook and Chris Everitt. (Midland, 1985).

aimer to see if he could find out what was going on. He too disappeared and again lost contact. I was beginning to feel a bit lonely in the cockpit and the screaming draught and creaking noises did not help.

'The aircraft had settled down and was handling well. The engines ran steadily. Eric Blanchard seemed happy with the fuel and thought the flying control rods were running smoothly so I asked him to see if he could find out what was happening in the rear of the aircraft making sure he had a duff oxygen bottle and our last feeble torch. He disappeared leaving myself at the controls and Howes feeling alone in the front. He was trying to ignore the excitement around him as he concentrated on his calculations, which were made more difficult because none of our equipment was working. Using dead reckoning to plot his Mercator chart he was trying to estimate our course for home and write up his navigation log.

'After what seemed an age, the intercom sprang to life.

'Engineer to Skipper, I've pulled the bomb aimer up the fuselage and am going back for the W/Op.'

'Heavy breathing and gasps were heard and then nothing. Another series of gasps as the microphone woke up again.'

'I've got the wireless operator on the bed and both of them on oxygen. Bit of a shambles at the back. I'll go and see what I can sort out.'

'The navigator looks out of the front windscreen trying to find a landmark. My eyes are covering every part of the sky searching for fighters and gazing below for flak.

'The mike came on again. 'Skipper from engineer.' I have plugged the mid-upper gunner into the oxygen. Cannot get the rear gunner to help himself. I am coming back.'

'My instincts were to dive down to a level where the air was less cold but light flak caused me to think again. I was not sure of the strength of the fuselage and I would need that height should we need to bail out or to fly home. Blanchard recovered enough to tell me what had happened. Only those who know how cramped and awkward the rear of a Lancaster is are able to appreciate his efforts. He had struggled in the dark and extreme cold trying to rescue four members of the crew. He was clad in his flying gear and relying on a bottle of oxygen. He said that we had a large hole nearly under the mid-upper gun turret. It seemed that Lamblin had gone to the rear and found that the wireless operator was in a heap on the floor. Sergeant K. D. Brown's foot had gone through a weakened part of the fuselage and was trapped by the slipstream and his foot was freezing. Lamblin had tried to free the trapped leg. Somehow his oxygen supply had failed and he was unconscious. After Blanchard had pulled Lamblin along the fuselage and plugged him into the oxygen he had then returned to rescue the wireless operator. He manoeuvred the W/Op's trapped leg through the floor and then got him on to the rest bed and plugged in the oxygen. He had tried to help the rear gunner but he was rigid in his turret.

'We lost height as soon as we safely could and flew back to Scampton, landed and taxied to the dispersal. I sat down and looked at my crew. I though 'gracious me, this is rough going'. The mid-upper gunner was trembling a bit and was very pale. As he took his helmet off there was frost between the helmet and the lining and it was beginning to tear the skin on his forehead. He had a cigarette. Lamblin came out looking shaky indeed but managed to say he was all right. He had a cigarette and sat down smoking. He would recover. Eventually the others emerged. The poor rear gunner was in a really dreadful state, trembling and hardly able to get a cigarette in his mouth. I was the only one of the crew who didn't smoke. An ambulance arrived and collected the wireless operator. It was discovered that he had a frostbitten foot.

'I got out of the aircraft feeling tired and drained. The rear gunner's face was ghastly and his hands trembled so that he had to make several efforts to light his cigarette. He could not control his lips when he did get the cigarette between them. Sergeant Martin had a slight sheen on his face. There were patches where the sweat had frozen under his helmet. Howes put down his flight bag containing his charts and equipment and smoked quietly and said, 'that was a bit fraught'. 'It certainly was! He had shown great skill in getting us back to base with no navigational aids working.

'Eric Blanchard had a look at the damage and helped me to fill in a bit of the Form 700, which included statements of damage and then lit his briar pipe and looked no more concerned than if we had finished a practice flight. He had been moving around in the dark, damaged and bitterly cold aircraft, attending to the injured and unconscious crew yet, as always, retained an air of calm confidence. Having finished his tour with me, he then went to train others at a Lancaster Conversion Unit and later he returned to the Squadron as Engineer Leader and was awarded a much overdue DFC. Sergeant K. D. Brown returned to us as soon as he had recovered from his frostbite. The rear gunner became quiet and withdrawn and often was too scared to give instructions from his turret. I kept him flying with me until his first trip to Berlin when he was literally scared stiff so he left us. The gunnery Leader, Flight Lieutenant Tony Wynyard flew two or three trips with me and Sergeant John George Smith, (or 'Smith 132' as he was known after the last three numbers of his service number), not yet 19, took over as rear gunner. He was very keen. As for me, I offered a prayer of thanks for safely escaping a dangerous situation and for having completed one more 'op'. Could our experience and luck hold out to complete a tour?'

Over 280 Lancasters of 1 and 5 Groups devastated the north-western and south-western sections of the city and a further 48,000 people were bombed out, making a total of 350,000 people losing their homes during the series of three raids in a week. Then it was Gelsenkirchen on 9/10 July. Thirteen 'Musical Mosquitoes' again marked with 'Oboe' sky-marking but the equipment failed to operate in five of the Mosquitoes and a sixth Mosquito dropped sky-markers in error ten miles north of the target and the raid was not successful. Seven Halifaxes and five Lancasters went missing from a force of 418 aircraft. 'Steve' Stevens noted in his log book that the flak was 'intense'.

On the night of 12/13 July 'Steve's crew were one of 295 Lancaster crews in 1, 5 and 8 Groups that flew to Italy to bomb Turin. 'Turin was beyond the range really, because my aircraft did barely one air mile to the gallon and it would hold 2,154 gallons so at that consumption you could so about 2,000 miles."

The main weight of the raid fell just north of the centre of the city. Over the target, conditions were excellent with no cloud and very good visibility; so much so that one crew commented on the novel experience of being able to clearly identify features of the town layout. The crews were unanimous in praising the PFF technique, the only criticism being that it was slightly late in starting. Visual identification of the two rivers and the town confirmed the accuracy of the attack which seems to have gone very well right from the commencement when marking and bombing was particularly described as both accurate and concentrated. As the attack developed numerous fires were observed and although in the later stages some scatter had developed, large areas in the North of the town and in the triangle formed by the Po and Dora rivers were reported as being a mass of fire with much black smoke up to very great heights. Large explosions were reported and the two biggest of these accompanied by bursts of flame were at 0157 and 0200 hours. Defences in the target area were not very troublesome and one crew in 1 Group described them as 'puerile'. The heavy flak was stated to have ceased about halfway through the attack. There was a moderate amount of light flak and also fifty searchlights whose operation seemed to have been particularly haphazard.

Stevens had looked in on 44 Squadron at Dunholme Lodge that lunch time and had met Squadron Leader [Lawrence William] 'Pil' Pilgrim and Flight Lieutenant John Sidney Shorthouse DFC; both of whom he had known for some time. "Pilgrim had once alleged that one night while he was trying to get away from a flak barrage he did a tremendous dive and realized he had lost too much height so he pulled the stick back and actually did a loop! I'd never known anybody else to do this. I would certainly never try it myself! Pilgrim told me to make sure I was on 'ops tonight as Squadron Leader Nettleton VC was flying. Shorthouse had taken a chap named Flight Sergeant Seager as his navigator. Seager told me he was flying as Nettleton's navigator. I had invited Seager to be my navigator while we were at OTU but Shorthouse had already snaffled him."

One of three brothers who all served in the Royal New Zealand Air Force during WWII, John Shorthouse was born in Portsmouth on 11 April 1920. In April 1939 he trained in New Zealand for entry on short-service commission to the Royal Air Force. On arrival in England he joined the RAF and in April 1940 following training on 207 Squadron was posted to 12 Squadron in France with the Advanced Air Striking Force. On 13 June 1940 he was shot down in flames and badly burnt. He bailed out and after two days was picked up by a Belgian ambulance driver, taken to Sainte Nazaire and repatriated to England. After recuperation he was posted to the Photographic Reconnaissance Unit at Heston and flew Spitfires on high-level photography sorties during the

Battle of Britain. Recovered, he spent a year in Canada instructing on Harvards and torpedo bombers.[12]

Maureen Miller was on duty in the control tower during the early hours of 13 July, listening out for the Lancasters returning from Turin and one voice in particular. She was always strangely convinced that 'Steve' would emerge, unscathed from the skies. 'His Lancaster was late and the control tower officer was worried, but I said, 'He'll be all right.' And he was.' Low on fuel, Stevens had put Lancaster ED944 down at Davidstow Moor on return. 'The long flight to Turin (10 hours, 10 minutes) which took longer getting there than getting back because you had to gain height first of all, meant that many aircraft could not return to base and had to be serviced and repaired so we had a longish break and I didn't go on another op then until 24 July.

Flight Sergeant Johnny Pickett a 20-year-old New Zealander from Mission Bay in Auckland who piloted Lancaster ED861 on the Turin raid was missing. 'We often discussed tactics before takeoff' Stevens recalled. "The previous day we had attended a service to dedicate the new chapel at Scampton." It was to be his last. His crew was the only loss on the squadron and Stevens would miss 'his bright intelligence and conversations greatly.' All seven crew members have no known grave and are commemorated on the Runnymede Memorial. In all, thirteen Lancasters, including 'Z-for Zebra' flown by Squadron Leader John Dering Nettleton VC were lost on Turin. Nettleton was shot down by a German night-fighter over the English Channel on his return. All seven crew including 21-year-old Flight Sergeant Dennis Ernest Arthur Seager who was from Lewisham, were lost without trace; their names added to those on the Runnymede Memorial.

Crews were not supposed to tempt fate and talk about death was taboo. In RAF parlance crews went 'missing'. At least it gave loved ones a glimmer of hope. 'Missing, believed killed' did not. In aircrew circles 'missing' were those who had 'Gone for a Burton' (a beer) or had 'bought it' or had got the chop; never 'killed'. Very seldom did crews, amongst themselves, say that they were on 'operations'. It was either 'dicing with

---

12. In October 1943 he was awarded the DFC and promoted to Squadron Leader and Flight Commander and in December 1944 Mentioned-in-Despatches. On 1 July 1945 he transferred to Royal New Zealand Air Force and in September 1946 he was repatriated to New Zealand.

death', 'juggling with Jesus' or 'gambling with God', reduced in most cases to 'dicing', 'juggling' or 'gambling'. Even so, bets would often be taken on who was going to be next to 'get the chop' or 'go for a Burton'.

'Steve' Stevens had been right about 'Dutchy' Haye and when bets were taken that an American pilot named 1st Lieutenant Don West would not 'make it' he was again proved correct. Born on 14 August 1918 at Fresno, California, West had joined the Royal Canadian Air Force on 27 August 1941 and he entered service at East Kirkby on 7 September 1943. On the night of 3/4 November he took 'P-Peter' off at 1722 hours on the operation on Düsseldorf and was shot down by a German night fighter while outbound to the target near Mönchengladbach and subsequently crashed at Hechtel-Eksel, Belgium. Flying Officer Robert Sinclair Clements RCAF the 2nd pilot (who was only along for a familiarisation flight) and Flying Officer James McPhail Elliott, the 22-year-old bomb aimer managed to evade capture. Pilot Officer Norman Buggy the navigator was taken prisoner. The crew members that were killed in the crash are buried in Heverlee cemetery. West was subsequently award a posthumous DFC. The raid also marked the first large scale test of the G-H blind bombing device used by 38 Lancaster Mk. II aircraft from 3 and 6 Groups, against the Mannesmann tubular steel works. The Gee-H or G-H radio navigation system replaced the Oboe bombing system which worked along similar lines. By measuring and keeping a fixed distance to a radio station, the bomber could navigate along an arc in the sky. The bombs were dropped when they reached a set distance from a second station. Oboe used very large displays in ground stations to take very accurate measurements but could only direct one aircraft at a time. Gee-H used much smaller gear on the aircraft and although not as accurate, could direct as many as 80 aircraft at a time.

During the summer months of 1943 'Steve' and his crew lived an almost charmed life. They carried 11,300 lbs of bombs on the raid on Hamburg on the night of 24/25 July when 'Bomber' Harris launched the first of four raids code-named 'Gomorrah' on the port. 'The briefing for this series of raids was quite amazing and brutally succinct; the object of this exercise is to destroy Hamburg!' Station Commanders were authorised to tell the crews that, 'Tonight you are going to use a new and simple counter-measure called 'Window' to protect yourselves against the German defence system." 'Window' consisted of packets of strips of black paper with aluminium foil stuck to one side and cut to a length (30 cm by 1.5 cm), which when dropped in bundles of a thousand at a time at one-minute intervals produced almost the same reactions on Würzburg ground and Lichtenstein airborne interception, radar as did aircraft. 'Window' was carried on the 791 aircraft – 347 Lancasters, 246 Halifaxes, 125 Stirlings and 73 Wellingtons – which set out for Hamburg. Just twelve bombers were lost in action; four Halifaxes and four Lancasters, a Wellington and three Stirlings. 'Good trip' was all that 'Steve' Stevens wrote in his logbook. "We needed to go to Hamburg on three successive nights at least but on the second night, 25/26 July bad weather over north Germany prevented all but a handful of Mosquitoes bombing Hamburg."

'Window' was still effective and so 705 heavies were dispatched to Essen instead. 'Steve' made a transcription of the wire recording of the raid for public broadcast.[13]

"During the day aircraft of 'A', 'B' and 'C' Flights are dispersed over the airfield. They have been thoroughly tested – guns, ammunition, bomb sights, compasses, engines, wirelesses, oxygen supplies, petrol and oil have all been carefully inspected. Tractors have drawn long lines of trolleys to the bombers. Each trailer carries one or more bombs and each bomb has been carefully hoisted into its correct position in the bomb bay. Buses now travel from aircraft to aircraft, and at each a crew is dropped off.

"Each of us is clad for our own particular job and each carries a bulky parachute. My clothing consists of clean silk and wool underwear – clean, because a piece of shrapnel or bullet may plaster my clothing into a wound, the cleaner therefore the better; a yellow shirt, open at the neck with a silk scarf inside. I wear a yellow shirt as I may be shot down over enemy territory and an Air Force blue shirt would be conspicuous and attract attention if I were to try to escape. I do not wear a tie in case I have to crash land in the sea. A collar and tie might shrink and cause discomfort if it didn't actually strangle me.

"Around my neck is a loose loop of string which has two plastic discs on it. Each bears my name, number, Air Force rank and blood group. If I am

---

13. Wire recording or magnetic wire recording was the first magnetic recording technology, an analogue type of audio storage in which a magnetic recording is made on a thin steel wire. The first crude magnetic recorder was invented in 1898 by Valdemar Poulsen. *'Hear It now'* with Edward R. Murrow was the first network radio programme produced and edited on wire recorders. It ran on CBS from 1950–51.

Lancasters on 97 Sqn practising low-level flying in the days before the raid on the M.A.N. U-boat engine plant Augsburg on 17 April 1942.

S/L John Dering Nettleton commanding 44 (Rhodesia) Sqn (2nd from left on the front row) and his crew who led the raid on Augsburg. P/O Patrick Arthur Dorehill, his second pilot is far left, front row.

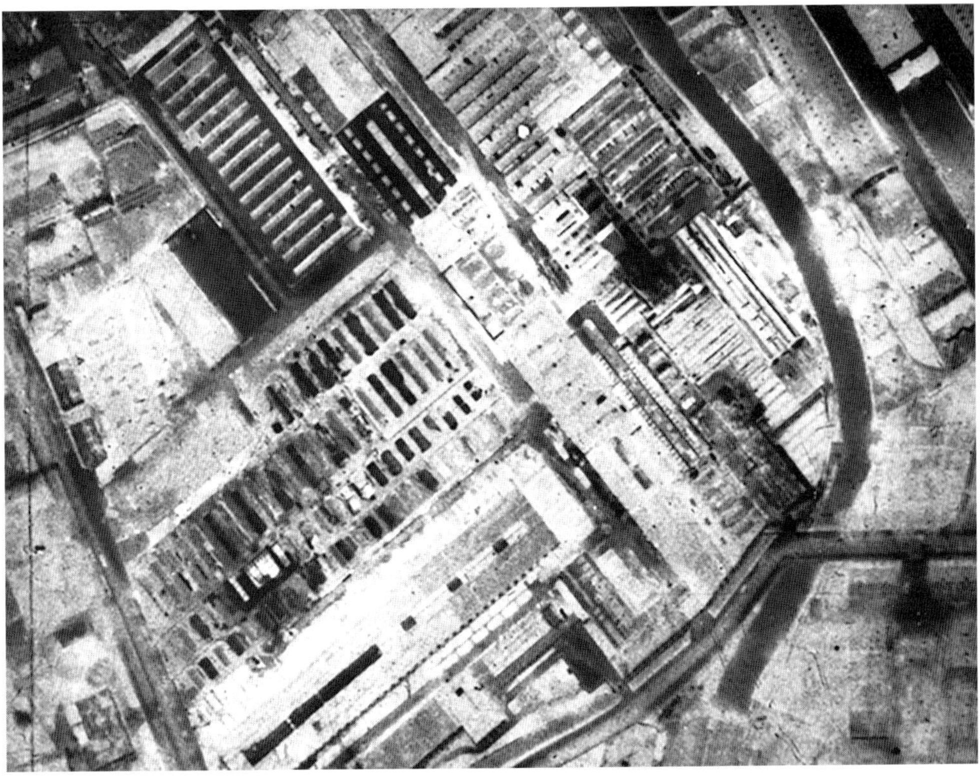

Reconnaissance photo of the M.A.N. works at Augsburg.

F/O Ernest Alfred 'The Devil' Deverill DFM and the crew of 'Y-Yorker' after their return from Augsburg.

Augsburg raid survivors pictured with Brendan Bracken the Minister of Information. Nettleton is far right, Flying Officer Brian Roger Wakefield 'Darkie' Hallows, 3rd from right.

The cockpit area of Lancaster R5702, VN-S "*Taipo*" (the Maori word for 'devil') on 50 Sqn that was damaged by flak on the raid on Hamburg on the night of 9/10 November 1942. S/L Roy Oldfield Calvert DFC** MiD who was injured, crash landed the Lancaster at Bradwell Bay. Two years later, on 15/16 February 1944 R5702 was shot down over Denmark by Oblt Gerhardt Raht of 4./NJG 3. Sgt William Ashurst and five of his crew were killed. Air bomber F/O Harry Proskurniak RCAF who had tried in vain to open the forward escape hatch was blown out of the fuselage. He came too while hanging in his parachute and was subsequently taken into captivity. (*Dave Homewood via Mrs May Calvert*)

The uninjured members of the crew of '*Taipo*' following the raid on the night of 9/10 November 1942, sit in one of the holes put through the wing by enemy action. Left to right are American P/O Power, 2nd pilot; Sgt Cruickshank, rear gunner; Sgts' Wilson, air bomber and Alan Connor RAAF. (*Dave Homewood via Mrs May Calvert*)

Lancaster W4842 '*Fema Dora*' that was so skilfully ditched by Sgt Ted Robbins on 27/28 May 1943 on the operation to bomb the armaments factory of Krupps at Essen.

Sgt Ted Robbins outside his parents' house in Cheshunt, Hertfordshire in 1942.

Leslie 'Cal' Calvert, the 30-year old bomb aimer on Sgt Ted Robbins' crew on 106 Sqn at Syerston.

Sheffield during the blitz on the nights of 12/13 and 14/15 December 1940. (New York Times, *Paris Bureau Coll*)

Wearing body armour 26-year old South African Pilot Officer John Laurens and crew, known on 101 Sqn as the 'League of Nations' crew, prepare to board ABC-equipped Lancaster DV267 SR-K at Ludford Magna prior to the night's operation. This aircraft and crew were shot down on 19/20 February 1944 on the raid on Leipzig by a Messerschmitt Bf 110. Laurens and two of the eight man crew were killed.

Flt Lt Andy Bray, AVM "Bob" Blucke DSO AFC (who had commanded Ludford Magna from 1943–45 and earned his DSO flying a 101 Sqn Lancaster on a raid against Mannheim in September 1943) and Association secretary Jack Morley at the Ludford Magna 101 Sqn Memorial in 1984.

P/O 'Steve' Stevens and fellow trainee pilot Len Gray with Alastair Cooke during a break from flying training at No. 2 BFTS, Lancaster, California in 1942–43.

Lancaster touching down at Waddington after another night op.

Essen under attack.

Flt Lt 'Steve' Stevens and Cpl Maureen Miller WAAF on their wedding day on 4 December 1943.

At left, S/L Richard J. Manton (KIA 20.10.43) with Group Captain John Searby (centre) and S/L Shaw on 83 (PFF) Sqn.

Lancaster ED421 UV-U on 460 Sqn RAAF which was MIA on Berlin 23/24 August 1943. Three crew members survived when the aircraft exploded. Those who died are buried in the Berlin 1939–1945 War Cemetery.

P/O Robert Hamilton Watts RAAF, 23-year old pilot on 'B-Baker' on 44 Sqn who was lost without trace returning from Leipzig on 20/21 October 1943 when he was forced to ditch his Lancaster in the North Sea.

Leipzig in ruins at the end of the war.

P/O Samuel Cunningham Atcheson DFC and crew on Lancaster JB474 on 57 Sqn who were shot down on the Stuttgart raid on the night of 15/16 March 1944. Standing L-R: Sgt Reece; P/Os' Atcheson and Antony Patrick McCall. Front: F/Sgt Jack Greenhalgh and Sgt Frank Slater Weaver. Not shown are Sergeants' Brian Henry Maude Thomas and James Willie Naylor. All the crew except Sergeant Reece were killed.

Sqn Leader Raymond Backwell-Smith, pilot of 'H-Harry' on 9 Sqn at RAF Bardney on 15/16 March 1944. (*Jerry and Rosemary Nice*)

Sgt Norman Sirman, flight engineer, 'H-Harry' on 9 Sqn on 15/16 March 1944.

Flak fills the sky over Stuttgart.

Stuttgart in ruins at the end of the war.

Flt Lt Surender L. 'Sammy' Berry RAFVR a navigator on 622 Sqn at Mildenhall was in hospital with tuberculosis when on the night of 27/28 April 1944 his pilot, Flight Lieutenant James Andrew Watson RCAF [below] was killed when ND781/GI-R was shot down by Ju 88 night fighters and Watson had remained at the controls of the stricken Lancaster while his crew parachuted to safety. He was posthumously awarded a MiD.

Crew of Lancaster LM121 on XV Sqn RAF skippered by Peter Charles Lewis D'Ombrain RAAF. Back row, from left: Leslie Arnold Hadder; Leonard Thomas Gearing; D'Ombrain; F/Sgt Arthur Stephen Long RAAF. Front row from left: F/Sgt Frank Bruce Reid RAAF; Sgt Raymond Geoffrey Norris; F/Sgt Laurence Seymour Jamieson RNZAF. Gearing, D'Ombrain, Long, Reid, Norris, Jamieson and F/Sgt Stanley Arthur Nystrom RAAF the replacement gunner were shot down and killed on 1 June 1944.

Lancasters of 463 Sqn RAAF in the snow at Waddington in the winter of 1943–44.

Lancaster taking off for a night raid on Germany.

KB700 *Ruhr Express* the first Canadian-built Lancaster X to arrive in Britain which arrived at Northolt, Middlesex on 15 September 1943 and was initially delivered to 405 'Vancouver' Sqn RCAF, but soon went on to 419 'Moose' Sqn RCAF at Middleton St. George, where it was eventually destroyed in a landing accident returning from Nuremberg, its 49th operational sortie, on 3 January 1945.

killed, one will be buried with my body and the other sent to England. If I am seriously wounded and need a blood transfusion, the doctor will know which group of blood to put into me. I am wearing battledress and the blouse of the battledress has many pockets. In one of these, I have a small escape outfit; in another, currencies of the countries over which I shall be flying. The contents of the escape outfit are interesting, and all will be useful to help me escape if I were to bail out over enemy territory.

"I have a pair of sheepskin-lined flying boots. In the leg of one, is concealed a knife so that I could cut the pieces covering my legs away from the lower part and change my flying boots into shoes. Again, this would be a help if I were to attempt to escape. Over the top of my battledress I wear a light windproof flying suit. I have a heavier one, too, but would only wear it if the cockpit heating system broke down. Over my flying suit is strapped a bright yellow buoyancy jacket which we call a Mae West. By pulling a lever, I could blow up this jacket so that I should be able to float for some time if I came down into the sea. Over the Mae West is a parachute harness to which I could clip my parachute if I needed it. I also have a soft leather flying helmet with combined oxygen mask and microphone and a pair of goggles. These goggles are secured to the top of my helmet and I should only pull them over my eyes if the windscreen were shattered or if the aircraft caught fire. I also wear long gauntlet gloves. The object of having long gauntlet gloves is to protect my wrists if the aircraft catches fire.

"The engineer, wireless operator and navigator are similarly dressed. The mid-upper gunner and rear gunner are much more heavily dressed. They wear huge electrically heated flying suits with electrically heated socks and gloves. Their turrets are not heated and they will be working in temperatures far below freezing for hours tonight.

"We enter 'E-Easy' and check everything in it with great care. The navigator sets out his charts and equipment. The gunners, engineer and wireless operator make sure they have everything in order. We stow away our carrier pigeons, too! I notice that there is a tiny spot on the windscreen and get it cleaned off. I could easily imagine that spot to be an enemy aircraft later in the night and I don't want to take unnecessary avoiding action from imaginary aircraft.

"Having checked everything, we get out of the aircraft again for a last stretching of limbs. A senior officer and a doctor come round for a last-minute check and then we go back into the aircraft. I adjust the seat and rudder pedals and make sure my equipment is comfortable. Then I fasten my safety belt tightly, clip on my face mask and switch on the oxygen. I call up each of the crew in turn to make sure they are comfortable and then tell my flight engineer we are ready to start up the engines.

"The cockpit is about twenty feet above the ground so the engineer has to signal the ground crew below with a light. Each of the four Rolls Royce engines bursts into life. The instruments quiver and the needles move on. The gunners rotate their turrets and each crew member again reports that all is well. I flash a light and the huge chocks holding the wheels are tugged away. I release the brakes and the aircraft – about thirty tons of it – taxies slowly under my control. I taxi steadily. I have 11,600 lbs of bombs – very nearly five tons – and thousands of rounds of ammunition as well as 1,850 gallons of petrol on board.

"At the point where we turn onto the runway, we stopped. The green light flashes from the caravan saying that I am the next to takeoff. We do not use the wireless to talk to the controller, or the Germans would hear us. Well, I suppose the Germans wouldn't literally hear us, but a number of transmissions would give them a warning that something was happening on our side of the Channel, and we want to take the Germans by surprise if we possibly can.

"I turn onto the runway, making sure that the aircraft is pointing right down the middle. Then I say to the engineer, 'Flaps Twenty, engineer' and he replies, 'Flaps twenty Skipper' and moves a lever.

"I then say, 'Crew, prepare for take-off' and slowly open the throttle whilst keeping the brakes applied. I release the brakes and we speed down the darkening runway.

"As the tail rises, the aircraft swings slightly and this I immediately correct with use of rudder and throttle. The speed builds up. I ease the control column back gently and we climb, ponderously at first.

'Wheels up, engineer.'

'Wheels up, Skipper.'

"Green lights in front of me change to red and then the red lights go out showing the wheels are now tucked up inside the engine recesses.

'Flaps up, engineer.'

'Flaps up, Skipper.'

"The engineer moves a handle and the wing flaps rise.

"Plus 9, 28, 50, engineer.'

"Plus 9, 28, 50, Skipper.'

"The deafening roar of the Merlin engines, each of which has been producing about 1,390 horsepower, dies down a little as the aircraft settles into a steady climb.

"The navigator then says: 'Navigator to Skipper; course Zero Nine Two." I move the controls, turning the aircraft gently until the compass reads 092.

'On course Zero Nine Two, navigator.'

"Then all is quiet except for the steady beat of engines. No-one will use his microphone now, except to pass a message. When a microphone is switched on, there is a noise which makes every crew member even more alert.

"Tonight we are going to bomb the Krupps works at Essen, but first we have to climb over England to a height of 20,000 ft. In fact, I have to be over Sheringham at 20,000 ft at 2202 hours to set course for Essen. At 2150 I say to the navigator, 'Skipper to navigator, height now 20,300 ft,' and to the engineer, 'Cruising revs and boost.' The engines again alter their note. I lower the nose of the aircraft slightly to gain a little speed and lose a little height and at 2159 the bomb aimer, who is sitting at the guns in front of the aircraft, says, 'Coastline ahead. It could be Sheringham.' The navigator gives a correction of course and we head over the sea to Holland and Germany. Every member of the crew is now very alert.

"The navigator and wireless operator are in lightproof compartments and cannot see out. Every other crew member has an area of sky to watch and this he will watch most intently. There are no lights other than those used now by the navigator and wireless operator inside the aircraft. Only the luminous pointers on our instruments show what is happening and where we are going. We are very aware that at any moment a fighter may come up from the dark shadow beneath the aircraft, pull up its nose and spray our underside with cannon shell and incendiary bullets. One burst on our bomb load and petrol tanks must mean immediate death for us. To allow the rear gunner to see as far underneath as possible, I bank the aircraft from side to side so that he can swing his turret and look underneath the aircraft. We have found that however much the Perspex turret is polished it still hinders visibility, so a large panel has been taken out. This gives the gunner a clear look but it means he is bitterly cold.

"When we are about five miles from the coast I give the instruction 'Gunners, test guns. The rear gunner leaves his microphone switched on so that the crew can hear the heartening noise of four Browning guns firing simultaneously. The mid-upper then fires his twin machine guns and the bomb aimer the front guns. Every fifth bullet is tracer so that a red stream seems to pour from each gun into space.

"The rear gunner reports, 'Rear guns OK, Skipper.' The mid-upper gunner reports: 'My port gun is misfiring.' He strips it down in the dark without using a light and then he says, 'mid-upper to Skipper, port gun has breech block trouble and I cannot repair it.' I don't like this and say, "Skipper to mid-upper, any suggestions?' There is a pause while the mid-upper gunner thinks, then he says, 'Could I change it for one of the front guns?' I reply, 'How long will that take?' He replies, 'About ten minutes, I think.'

"The decision is now mine. I think the chances of using the front guns are slight tonight as we are flying high. If we were flying low, we should need our front guns for strafing. On the other hand, if I immobilise the front and mid- upper turrets for ten minutes, this could be fatal if we were attacked. Again, German night fighters are more likely to be met after we cross the enemy coast than at this particular point. I therefore instruct the bomb aimer to dismount one gun from the front turret and the mid-upper gunner to dismount the faulty gun from his turret. The wireless operator gives the mid-upper gunner a hand. Before he can move, of course, he has to plug into an oxygen bottle. The mid-upper dismantles his gun. The wireless operator sits in the mid-upper turret keeping watch while the mid turret gun is taken to the front turret. The mid-upper gunner passes his gun to the engineer who passes it to the bomb aimer. The bomb aimer puts this gun in the front turret and passes the good gun from his turret back to the mid-upper turret.

"Still working in the dark, the mid-upper gunner and the wireless operator between them assemble the gun in the mid-upper turret. We are relieved to hear the microphone click and the mid-upper gunner say, 'Guns remounted. Permission to test?'

"I say, 'Test guns and look out for fires." Both guns work well and we breathe a sigh of relief. Very few minutes have in fact passed since the first fault was discovered, but it seems quite a long time to me. Minutes later, the bomb aimer calls; 'Enemy coast ahead!' He has seen the coastline of Holland.

"As soon as our aircraft have reached 15,000 to 20,000 ft, the German radar operators have been watching us. At first they have picked us up on Freya. This equipment has a range of about 90 miles and gives a bearing and distance of an aircraft, but not its height. As our bombers approach the enemy coast, a second group of radar instruments known as Würzburg takes over. They have a range of about 45 miles and can give height, speed and bearings of aircraft very quickly. This information is given to German night fighters and flak regiments. Now the German squadrons of night fighters warm up their engines and check fuel and ammunition. Searchlight operators and flak batteries stand by. The battle is about to begin.

"The enemy coast has been crossed. We see some searchlights probing in our direction and see some bursts of anti-aircraft fire. Usually we take little action about flak unless we can hear it crack or smell it. We have to get to a target on time and we don't want to dart about all over the sky unless there is real danger.

"A single searchlight picks up an aircraft on our starboard side. Within a second or two, a huge number of searchlights illuminate this aircraft and gunfire flashes around it as it twists and turns in an effort to escape.

"We know that every gun in the area is directed at this aircraft. At first it seems that it may have escaped but suddenly there is a brilliant flash; a red glow and a ball of smoke, and we know that at least seven of our men have been blown to pieces.

"The searchlights flicker again and catch another aircraft. Suddenly, a series of flashes ahead makes me jump. I realise a fighter, quite invisible until that moment, has tried a head-on shot at us, with cannon and machine gun. He speeds by overhead, narrowly missing my aircraft. This happens so quickly I cannot report it. The gunners have seen the flash and the mid-upper shoots at the aircraft but it has gone away, disappeared. We see flashes of tracer as other fighters shoot at other bombers, and then the sky seems empty and we drone on.

"As we approach the Ruhr Valley we see searchlights ahead of us; thousands of angry red flashes of bursting shells appear, leaving behind dark mushrooms of smoke. I know that before long I shall have to fly my aircraft through those searchlights and through the gunfire to get to the target.

"I check each needle on every instrument carefully and ask the engineer to switch to the fullest tanks. The navigator gives the wind speed and direction to the bomb aimer, who sets it on his bomb sight. I turn up the oxygen so that we have a maximum supply. The rear gunner cracks the ice which is forming on the inside of his oxygen tube. The wireless operator has been tuning in to the wavelengths used by German night fighters. If he hears a German voice he pushes a key and blasts out the noise of our engines and the wind blowing past us. This, we hope, will upset the wireless communication between the German night fighters and their ground control.

"Now we are among the searchlights. Anti-aircraft gunfire bursts on each side. Night fighters blast in firing at bombers and bombers shoot back. Our bomber suddenly lurches and drops several hundred feet as a shell bursts nearby.

"If you could see the navigator at this moment, you would find him rescuing his instruments from mid-air. So far, we have not been hit and only had one small attack. Suddenly, a stream of red and green markers burned brilliantly on the ground. These are target indicators dropped by Mosquito bombers flying much higher than we are.

"I say to the bomb aimer, 'Aim for the red target indicator.' At this point the target disappears under the nose. The bomb aimer, lying in the nose, sees through a transparent panel all the target area set out in front of him. He lines the red target indicator up on his bomb sight and operates the switch to fuse the bombs. He knows I hate to open the bomb doors too soon as this makes the aircraft less manoeuvrable. As the target appears, he gives me instructions: "Left – left – right – right – steady – steady – steady – left – left – steady – steady.' Then he says, 'Bomb doors!' I move a lever and say, 'Bomb doors open.'

"Steady – steady – steady – steady,' says his voice and I concentrate on holding the aircraft level and exactly on course. This is a most trying time. If a fighter comes in, we cannot take avoiding action now. An explosion rocks us. An aircraft nearby is blown up.

"Right! Right! Right! Steady – steady – steady,' says the bomb aimer and I put every mental and physical effort into carrying out his orders.

"Suddenly I feel the aircraft lift and know our bombs are dropping. Then comes the voice of the bomb aimer, 'Bombs gone!' He strains his eyes to see the results. With the bombs we drop a three-million candle power photo flash which burns dazzlingly, enabling a camera we have in the aircraft to take photographs of our bombing.

"I fly straight and level for another thousand years – only thirty seconds or so as we measure time now, while the camera records its evidence. Then I open the throttle and a lighter aircraft bounds away.

"We set course for home, knowing that between us and England there are still fighters, anti-aircraft guns and searchlights and our task is not yet over.

"However, this time we've bombed on the first run. On one occasion we had to fly over the target three times to ensure a hit. Even after this period of time I still get goose pimples over the thought of that one.

"As we settle down on course for home, the air seems suddenly empty. The target glows behind us and we still find an odd searchlight flicking in our direction but nothing else disturbs us. The earth is black beneath, stars wink above and I see the Pole Star clearly, nearly on the starboard wingtip. We drone on, still banking, so that the gunners can look underneath.

"Suddenly, we are galvanised again. A microphone clicks. 'Tail to Skipper, fighter, fighter, port quarter down, range 600, port, port, go!' I push on the rudder, turn the ailerons, put the nose down, open the throttle and transmit to the gunner, 'Diving port!' I dive for the dark part of the sky and as I'm going down port, the gunners aim for the upper starboard side of the fighter.

"The fighter blasts a frightening burst of cannon and machine gun fire, some of which I can feel hitting my aircraft. A stream of red and yellow fire passes me and I seem to be in the middle of it.

"The mid-upper gunner says: "Fighters broken to starboard up" and I bring the aircraft back onto its course for home. The German fighter pilot is a persistent fellow, however, and again I hear, 'Fighter, fighter, starboard down, range three hundred! This time I say, 'Corkscrewing.'

The gunner says, 'Starboard, go!' and I'm surprise how calm our voices are.

"I say, 'Corkscrewing down starboard.' The gunners aim above the fighter's port wing and fire long bursts. I don't want to be driven too low, so next I say, 'Changing.' The gunners then aim their guns point blank at the approaching fighter.

"Then I hear *'Hit! Hit! Hit! Hit*!' The guns continue to hammer and then stop. 'Look, Skipper, he's hit; port down! He's burning!' I look and see a red glow in the sky and a yellow flash and I find my voice, saying, 'Switch off your microphones, keep searching.'

"The navigator gives me a change of course, and says, 'Dutch coast, nineteen minutes.'' We fly on, every minute seeming like an hour. Looking out towards the wingtip, I see a mist breaking over the wing. The engine temperature of the starboard engine is rising slightly and I suspect the coolant is leaking. I ask the engineer to keep a careful eye on the radiator temperature of the starboard outer engine.

"We continue to fly and once more on our left we see this terrible picture of an aircraft suddenly being coned and then shot down. Then the enemy coast passes beneath us and we are over the sea. The engines feel rough and laboured and we gently lose height to 10,000 ft.

"Suddenly the engineer says, 'Fire starboard outer!' and I see the engine glowing red. He cuts off the petrol while I haul back on the throttle. We wait until the propeller starts to windmill and then he pushes the 'feather' button. The propeller blades turn until they are facing fore and aft and the engine stops. Then the engineer pushes the fire extinguisher button.

"I find myself saying 'Skipper to crew; engine fire starboard outer, prepare to parachute.' The navigator quickly gives our position to the wireless operator who is ready to send an SOS. If that fire takes hold the wing may burn through in twenty seconds. During that time the wireless operator will give our position and clamp his key so that air sea rescue vessels know where to look for us. However, the fire dies, blanketed by foam from the extinguisher and we fly on, still banking, still watching, still searching.

"Getting near the coast of England, we switch on our IFF. This identifies us on the radar screen as friendly. We put a special cartridge in the Very pistol and fire it as we cross the coast, to make certain none of our own anti-aircraft gunners shoot at us. Then we fly to our own airfield.

"As I near the airfield, I switch on my radio and call, 'Hello 'Heron', 'Lounger Easy', over.' Maureen's friendly, warm voice comes back, 'Lounger Easy' from 'Heron', pancake!'

"I lower the wheels and flaps, line up with the runway lights and land. The Lancaster lands easily on three engines, a tribute to its designers. Then the crew goes to a room to fill in forms about the flight and tell the intelligence officers about the trip. Twenty-six aircraft are missing from the Essen operation. People bring us mugs of hot coffee and those who feel like it, eat breakfast. Two other crews saw a battle going on in the positions which we described and they confirmed that a German fighter appeared to have been shot down, so we went happily to bed.

"We had put forward any complaints or concerns about the engines or the handling and while we were debriefed the duty staff would begin working on the problems. The exhausts might distort because of overheating when they were breathing out flame, so the fitter or rigger would go along with a screwdriver and tap the exhausts and if there was a dull sound, they had to be changed. We had lots of ground crew who had been trained as boy entrants and they were very skilled. They could set their hands to almost anything successfully. The people I had as ground crew were absolutely first class. Flight Sergeant Mike Booth, fatherly figure almost like my father, always dapper, very smartly dressed, kept his crew well under control. He liked them and got on very well with them. I can't praise them too highly. They were badly paid, not very well accommodated and when they could, tried working 8 hour days, but that wasn't possible when we were flying that night, so they had a very rough time of it. You knew very well that you just needed someone to be careless tightening a bolt and you would be in trouble. Sometimes you just had to turn back when the aircraft was in such poor condition that you couldn't fly it and you knew you couldn't reach the target.

"The next night we might be flying again" wrote Stevens.[14] They were. On 27/28 July it was Hamburg again. "This was an amazing raid when we started the first firestorm. It was ferocious. I was up at 22,000 ft and one column of smoke was as high as I was. My notes say that it was a 'quiet trip'. The defences were so disorganized that we only lost seventeen aircraft. Only 29 aircraft shot down on the first two attacks on Hamburg. 'Window' was performing well. According to my calculations, we had lost 292 aircraft shot down over Germany and a good 30% had arrived

---

14. Twenty-six bombers (or 3.7% of the force) failed to return of which, nineteen were destroyed by night fighters. In all, 627 aircraft out of 705 despatched dropped 2,032 tons of bombs on Essen.

in this country damaged. In the twelve raids in June and July which I had flown, the average loss had been 12.3 aircraft per operation. I had been intercepted by night fighters three times, shot down one after a fight during which one of my engines was severely damaged, had minor flak damage except on one occasion when a large hole was caused in the underside of the fuselage, the gunners were temporarily unconscious because of fractured oxygen lines and the wireless operator was frost bitten. I then had my second nine day break. The end of two exciting months and felt I was more skilled in the art of war.'

In August Steve's' crew flew seven ops including two to Milan. "On one of the trips to Milan it was a perfect night; absolutely wonderful. You could see everything. You could see Switzerland. There was hardly a bump in the air. I went across another night and there was a hideous thunderstorm and then of course you find all sorts of things you've never experienced. I just sat in the aircraft with this dreadful noise going on outside and the rain was thundering against the fuselage and suddenly there was an enormous cracking noise so I said to my flight engineer, 'Go and see what that noise is.' He said 'It sounds like flak'. And I said, 'It can't be flak in a storm like this. He was sitting on the floor and getting bumped around all over the place. I suddenly had a thought. 'I know what this is! We've got ice formed on the propellers and as they're spinning round the ice is being thrown off against the fuselage." Icing on the propellers was not unusual. On this occasion the ice sounded like bits of shrapnel hitting the fuselage; a tremendous noise. I'd never experienced it before. You are fighting to hold the aircraft in a thunderstorm like that. You could drop a thousand feet almost vertically and then you've got to get control of the damned thing to get it up again.

'The other strange thing was static electricity. I found by some strange chance when I was pointing something out to the bomb aimer there was a spark going from my finger onto the windscreen. I thought it interesting. I could almost write my name on the windscreen. And of course the gunners had St. Elmo's fire. You got this over the props anyway. It was like a halo around the aircraft.'

On an op on Berlin on the 30th, Steve's' crew 'Boomeranged' (aborted) when oxygen shortage forced an early return and so their number of ops to date numbered twenty. They had another 'Boomerang' on 3 September when the target was the 'Big City'. Because of the high casualty rates among Halifax and Stirling aircraft in recent raids on the 'Big City', the raid would be made up entirely of 316 Lancasters. 'Steve's' crew suffered a series of technical problems which forced them to abort. The port outer engine went u/s, as did the mid-upper gunner's turret and he and the bomb aimer's heating also failed.

On 5/6 September the crew successfully completed their 21st operation with a trip to Mannheim. Flight Sergeant Wilson Yates, a former Preston police constable flew as second dickie.[15] Just over 600 crews took part in a double attack on Mannheim and Ludwigshafen. The aiming point was in the eastern half of Mannheim with an approach from the west. The raid was successful but 34 aircraft were lost.[16]

'Steve's next op was to Hannover, on 27 September, when he had to divert to Waterbeach on the return. The crew nevertheless chalked up their 22nd op. Two nights later, on 29 September, when the target was Bochum the navigator was taken ill and the 'op' was another 'Boomerang' for the crew. By now 20-year-old Flight Sergeant Richard William Newcomb had joined the crew as bomb aimer. Newcomb was from Olmué, in the Marga Province, Valparaíso Region in Chile. His mother Elena was Chilean and Clive, his father, British. "His English was perfect" said 'Steve' "and his clear voice gave me precise instructions when flying my Lancaster over the very heavily defended Ruhr Valley targets and others as scattered as Hamburg, Leipzig, Berlin, Milan and Turin. I never once heard him hesitate even when we were doing the often violent manoeuvres to avoid enemy fighters and flak."

Operations 23 to 28 for 'Steve' took place on six nights in October, in JB311 'B-Baker' to Hagen, Munich, Stuttgart, Hannover (twice) and finally, Leipzig, ninety miles southwest of Berlin, for the first serious raid on the city, on the night of 20/21 October when 358 Lancasters of 1, 5, 6 and 8 Groups were dispatched. The weather was described as 'appalling'. Just 271 aircraft bombed the correct area and the rest of the bombing was very scattered. About 220 night fighters were despatched, returning

---

15. Pilot Officer Wilson B. Yates DFC was killed in action on 630 Squadron on 20 February 1944.
16. Thirteen Halifaxes, 13 Lancasters and eight Stirlings, or 5.6% of the force.

with claims for eleven Lancasters destroyed for nine own losses. Sixteen Lancasters failed to return but JB311 returned safely.[17]

After the Leipzig raid, 'Steve' was tour expired. 'By the end of my tour, 669 of our aircraft had been shot down but I only knew this afterwards.' Three bomb aimers who had flown with him on occasion were among those lost. Two were killed and one was taken prisoner. Richard Newcomb, who had caught tonsillitis and missed some operations still had more ops to fly was killed on the Berlin raid on the night of 18/19 November 1943 when his Lancaster, piloted by 24-year-old Flight Lieutenant Anthony Francis Gobbie DFC was shot down at Bohnsdorf in the Treptow-Köpenick district of Berlin. The skipper and four of his crew were taken prisoner but Newcomb and the navigator, Pilot Officer Alfred Edward Walter Gardner DFC, were lost without trace. Newcomb had just been promoted to pilot officer and probably did not have time to buy his uniform. Flight Sergeant James George Louis Martin DFM was killed on his second tour on the night of 18/19 September 1944 on the raid on Bremerhaven with just one survivor on Flying Officer Robert John Waugh's crew. John George Smith was killed on the night of 23/24 November 1943 on 630 Squadron after his Lancaster took off from East Kirkby at 1706 hours for Berlin. His 20-year-old skipper, Pilot Officer Joseph Howe and the five other members of the crew also died. All are commemorated on the Runnymede Memorial to the Missing. In January 1944 the Germans confirmed the deaths of five members of the crew but made no mention of the pilot or the flight engineer. Flight Sergeant Eric Blanchard was commissioned and flew a second tour on 57 Squadron as Engineer Leader. He was awarded a well-deserved DFC and survived the war.

In November 1943, just days before his wedding, Flight Lieutenant Stevens shook hands with King George VI as the monarch awarded him the Distinguished Flying Cross. On 4 December 'Steve' and Maureen married at St. Matthew's Church, Thorpe Hamlet in Norwich. They were granted nine days' leave for a honeymoon in Torquay. 'Lord Nuffield paid for 'Steve's honeymoon because he was an RAF officer' Maureen recalled 'and when we went to pay my bill, someone had paid it. Who, I

---

17. Lancaster III JB311 was lost with Pilot Officer Keith Cumming McPhie DFC RAAF and crew on 29 January 1944.

never found out.[18] In 1945 I was expecting our son, Adrian and came out of the WAAF with £12 and 10s – it wasn't even enough to buy a pram!'

The couple lived with her parents in Norwich until Adrian was six. After her war service Maureen worked as a secretary for Colman's Mustard for thirty years. A pilot once recognized her and said, 'Hello Heron! 'Steve' carried on flying for the RAF after the war but also trained as a teacher and taught maths at schools in Norwich. He loved teaching. 'I found maths difficult at school so I found good ways of learning it because it was essential to flying' he said.

Maureen passed away in December 2017 aged 97 just before the couple's 74th wedding anniversary and 'Steve' Stevens died in April 2020 aged 98.

---

18. 'It is not generally known that about mid-1942, that generous charity the Nuffield Trust, in addition to the thousand and one other acts of beneficence which they sponsored, provided a special cash leave allowance for Bomber Command aircrew. It had transpired that, aircrew receiving a greater leave allowance than ground personnel, some aircrew felt they could not afford to take it, financially speaking and spent their leave on the Station. This was held to be bad for morale but there seemed to be no answer until the Nuffield Trust stepped into the breach. The rates varied according to category, but a Pilot Officer could look forward to thirteen shillings plus a few shillings more as flying pay. Out of that came domestic allotments and mess bills and therefore the extra pound or two for leave from the Nuffield Trust, was much appreciated.' Gunner's Moon by John Bushby (Futura Publications 1974).

Chapter 5

Rustle of Spring

In the evening of Wednesday, 20 October 1943 at about half past nine, 14 year-old Harry de Waard was watching with his parents, an aerial combat between a German night fighter and a Lancaster behind their house, Steendijk 118 at Anreep, a small hamlet near Assen in Holland. The Lancaster came down on fire. 'The fireball grew steadily as the Lancaster sped towards the earth' Harry said. He thought it might have crashed on or in the neighbourhood of the Licht en Kracht mental hospital. Later it appeared that the Lancaster had crashed in a meadow owned by a farmer near Eleveld. Because of the enormous blaze and the noise of exploding ammunition it seemed to Harry to be much closer. They could not go to the scene of the crash because after 2000 hours there was a curfew. The next day Harry went with a couple of friends to the meadow, but they were not allowed to approach the completely burned-out Lancaster. The police and German soldiers had fenced in the meadow. The Germans quickly took over the guard from the Dutch police and kept inquisitive people at bay. At first five crew-members were counted. Later the Germans found another body and still later, a seventh member of the crew. Their mortal remains were taken away and were buried in Zuiderbegraaf Plaats Cemetery at Assen. The local civilians were prevented from attending the funeral by the German occupiers and their dissent was made obvious the following day when the seven graves were found covered in flowers.

Leipzig, 99 miles southwest of Berlin and about 250 miles east of Kassel, in eastern Germany, was the target on the night of 20/21 October 1943. It meant a trip of seven hours and ten minutes. This was the first serious raid on the city. Once one of the major European centres of learning and culture in fields such as music and publishing, it became known as 'Little Paris'. At the outbreak of the war the city had more than 700,000 inhabitants. The Erla Maschinenwerk aircraft factory produced Messerschmitt Bf 109 fighters at the three locations of Heiterblick, Abtnaundorf and Mockau. Additionally, Leipzig was an important rail intersection in Germany at that time. Prior to 1942, Leipzig had been considered relatively safe from air attack because of the long flight route from Britain but on 27 March 1943, bombs dropped by Bomber Command set Gohlis, an area in the north-west of the city, on fire. On the night of 31 August/1 September, Bomber Command carried out minor attacks on the towns of Eutritzsch and Schönefeld causing four casualties.

At Little Snoring in North Norfolk in 3 Group 115 Squadron provided fourteen Hercules-radial engined Lancaster IIs for the Leipzig operation. By the end of the war this squadron would have the distinction of being the one with the most operational service, most losses by any one single unit and the most tonnage of explosives dropped. One of the Lancasters on the Battle Order was 'F for Freddie' skippered by 26-year-old Flight Lieutenant John Thomas Anderson of Shield Row, Stanley in County Durham. The second pilot, 27-year-old Ronald Cyril Saville Clements was married, to Gertrude Muriel, who was living at home in Sale, Cheshire. Sergeant Herbert Gilbert McDonald Batten the 24-year-old flight engineer was one of two brothers.[1] Flight Sergeant Derrick Leslie Walter Horn was the 20-year-old navigator. Flying Officer Douglas James Boston the 23-year-old wireless operator was from Greenford in Middlesex. Flying Officer Frank George Andrews the 20-year-old bomb aimer was from Bournemouth in Hampshire. Sergeant Ernest Alfred Gibbs the 22-year-old mid-upper gunner was from Winson Green in Birmingham. The rear gunner was 24-year-old Flight Sergeant Frank Norman Simpson Cowie.

It was late in the afternoon on 20 October that Squadron Leader Richard John Manton had taken 'A-Apple on 83 Path Finder Squadron off from RAF Warboys, a satellite airfield for Wyton in Huntingdonshire, whose runways were under repair. Manton, born in 1913 in Melbourne, Derbyshire was the son of Brigadier Lionel Manton of Dower House. He joined the RAF in 1937 and when the war broke out, he was stationed in India. He served throughout the Burma Campaign returning to England

---

1. Sergeant Harold Royston Batten, a WOp/AG on 161 Squadron at RAF Tempsford would be killed at age 21, on Operation 'Pennyfarthing' on 11 November 1943. Following engine failure the pilot of his Halifax lost control at 3,000 ft and crashed at Brunelles near Chartres in France killing all on board except for the air bomber, who was later captured while trying to escape to England.

in November 1942. He was invested with the DFC at Buckingham Palace on 21 October 1941. Walter R. Thompson DFC*, a Canadian pilot on 83 Squadron who also flew on the raid, wrote the following of him: 'Manton was an exceptional pianist who, at 30 was a little older than de group of young aircrew who frequently gathered around the piano and asked him to play. A sensitive man who, I suspect suffered from unrequited love, because he played mostly melancholy music. In the late afternoon before we took off we asked him to play *Rustle of Spring*, one of his favourite numbers, by Christian Sinding.[2] After he'd played the piece perfectly we then dressed and went to our planes.[3]

Manton's navigator was Squadron Leader Archibald George Alexander Cochrane DFC. 'Archie' Cochrane was born on 28 August 1911, the son of Percy Cochrane and Louisa Gordon. He had three younger brothers of whom only two would survive the war. His brother Gordon died of malaria in Rhodesia (now Zimbabwe). Archie enlisted in the Royal Air Force in 1934 and in January 1939 graduated as observer on 108 Squadron. In 1942 he became a navigator in 25 Group Air Training. He had married Iseult Sambridge in 1940 in Cambridge and when his son Ian was born on 11 May 1943, Archie was returning on a flight from South Africa. In June he joined 83 Pathfinder Squadron at Wyton and was assigned to the crew of Squadron Leader Manton. Archie Cochrane was a passionate photographer. Normally, he had his camera with him on every flight but on 20 October he left it at home.

Flight Sergeant Albert Charles Branch DFM the 21-year-old flight engineer was born in Aldeburgh, on Suffolk's east coast, the son of Charles and Ruby Branch who also brought up two other brothers and three sisters. Charles Branch had served in the Royal Flying Corps in 1917 and later in the RAF in World War Two. Albert and his two brothers were choirboys in Aldeburgh Parish Church during their teens. At fourteen Albert had begun his apprenticeship in the garage where his father worked as senior mechanic. At the outbreak of the Second World War, he had already left the garage, finding work at a nearby airfield. In November 1939 Albert volunteered for the RAF. After basic training he began training as a technician. He worked for many months in that position, until he was

---
2. Frühlingsrauschen, Op. 32, No. 3 or *Rustle of Spring* is a solo piano piece written by the Norwegian composer Christian Sinding (1856–1941) in 1896.
3. *Lancaster to Berlin* by Walter Thompson DFC and Bar (Goodall Publications Ltd, 1985).

selected to become a Flight Engineer. On 6 June 1942 Albert Branch was posted to 50 Squadron at Skellingthorpe, near Lincoln. In April 1943 he finished his first tour of 33 night operations and on 4 August he was recommended for the Distinguished Flying Medal for Meritorious Service by Group Captain Sam Elworthy the station commander and he received his award from the 5 Group AOC, Air Vice Marshal Ralph Cochrane. After a posting to 1661 OCU, Branch joined 83 Squadron on 25 September, where he flew his first operation to Hannover. There followed another three operations, to Munich, Stuttgart and Hannover. Leipzig was his 37th and final operation. At the time of his death Albert was engaged to be married. His mother and father lived to the ripe old age of 99 and 98 respectively and had been married for 78 years on the death of Charles in 1999.

Albert Edward Evans the 32-year-old wireless operator was born in 1911 in Hull, the youngest of three children. Although he never married, he had had a serious girlfriend during his training time in South Africa. Had he survived the war he would certainly have settled in that country. Clarence William Foster the 21-year-old rear gunner was born on 5 October 1922 in Aberdeen, Scotland. He was the only son of five children granted to his parents Clarence William Foster and Barbara Arthur. At the age of seventeen Clarence with two friends tried to enlist with the Royal Air Force but because their parents did not consent they had to wait another year before joining the RAF. Like so many young men who were regularly confronted with death, they looked for romance in the vicinity of the airfields where they were stationed. In 1942 in Doncaster to the north of Lincoln, Clarence met a girl with whom he became engaged. Although they had surely made plans for better times, they were never to be realised. Sergeant Frank Earnshaw the 22-year-old mid-upper gunner was born in Romford, Essex. During the flight to Leipzig and back, he sat on a small seat, often in bizarre circumstances. Lancasters had no heating and the crew wore electrically heated clothing, gloves and shoes. At temperatures sometimes as low as 40 degrees below zero, this was certainly no unnecessary luxury. Flight Sergeant Roy Leonard Taylor, the 20-year-old air bomber completed the crew. Leipzig would be his 39th operation.

'A-Apple' was one of 73 Pathfinders that led 313 other Lancasters of 1, 5, 6 and 8 Groups on the 1,200 mile journey to Leipzig. Just 271 aircraft bombed the correct area and the rest of the bombing was very scattered. The bombs were all dropped in 36 minutes and the journey home could begin. The weather was appalling with heavy thunderstorms, hail, snow and heavy clouds. Icing prevented high-altitude flying which probably sealed the fate several of the missing Lancasters. Some crews on 83 PFF Squadron had difficulty with icing and electrical storms. Two crews were forced to turn back with a variety of troubles ranging from compasses to instruments, to turrets. Two others were damaged by fighters; one, Pilot Officer J. S. Simpson lost half of his tail, his hydraulics, his air pressure, his starboard inner engine and had a burst tyre. He managed to land at Newmarket, on the race-course. Pilot Officer Britton suffered three attacks, losing part of his tailplane, one engine, his flaps and part of his mainplane. His rear turret was knocked out and his upper turret damaged on the first pass. He could not hold the control column far enough forward so he got some of the crew to help him hold it while his engineer, Flight Lieutenant Forster, went aft, found the elevator trim wires and adjusted them manually. Britton made a masterful landing with a burst tyre. The markers were dropped blind on H2S but that was not much use when most of the Main Force could not see them through the clouds and many of the bombs went astray.[4]

One Lancaster on 207 Squadron almost did not make it back. The crew was approaching the target when a Lancaster above them was attacked by a night-fighter. They saw the tracer bullets coming from above and behind. The aircraft above dropped its bombs and they were hit by one of its incendiaries, starting a fire in their aircraft. Then the night fighter turned his attention to them. Harry Sparks the wireless operator was able to see the attacking fighter close up: 'So he swooped down, fired at the aircraft above us, hit it, burned it, swung round and in that second suddenly found he was on top of us. He must have been fairly experienced because he flew straight down and seemed to draw another attack position without even thinking. His nose came round and, as he opened up, there were great lances of fire at our side; he was that close. Then he veered off because he was too near – we could almost have shook hands with him. It was all very fast, but he swung away to avoid being blown up with us!

---

4. *Lancaster To Berlin* by Walter Thompson DFC and Bar (Goodall Publications Ltd, 1985). Britton was able to make a safe landing at Hardwick near Wellingborough in Northamptonshire. *RAF Bomber Command Squadron Profiles.*

'I was hit. A shell came through the fuselage and hit me low, down through the top of my legs and it lifted me up and smashed me right across the soft edge of the structure. I fell down onto the floor, ending up underneath the navigator's table which was only a short distance away. I pulled myself up, because everything below my waist was in a hell of a pain. Because it was dark I didn't know how it had happened or really where I was. I'd got field dressings stuffed in the hollows between the ribs at the side of the aircraft so I grabbed one of them, ripped it open, and smacked it underneath my leg. Then I screwed it up tight, right into my groin like a tourniquet to stop the bleeding if it was the femoral artery. Lying on the rest bed I flicked on the intercom and called 'Ho Jimmy' (the mid-upper gunner) and 'Ho Dev' (the rear gunner) but there no replies from either of them, not a sound. I could hear the navigator but he and the pilot couldn't hear me because the intercom had gone u/s. The pilot needed to talk because the bomb-aimer was ready to let go of the bombs. It was a hell of a position to be in. I slid down under the navigator's table, pulled myself across and in the dark ran my fingers all the way along where any wires would be, to see if there were any breaks.

'I could hear the pilot yelling 'can you hear me? Can you hear me?' He had no communication. So I grabbed down and felt where the wiring was breaking up and, in the dark, got hold of it, and turned it all round. I was lying on the navigator's feet – he was a very big man.

'All of a sudden I heard the skipper's voice coming over, 'OK': I'd made contact! So I said, 'Hallo there,' and he replied, 'How are you?' and I said, 'I've been hit.' I dragged myself onto the seat and saw that Jimmy, the mid-upper gunner had also been hit. Then I realised we'd been hit by cannon fire.

Jimmy with his great yellow suit on was hanging out of the turret in the open with his head and helmet dangling down. All the perspex had gone and he was just hanging there in the fuselage. The cannon shell had come through the perspex. Because his head nearly touched the metal of the turret, the shell had come right across the top of it, in the centre, and punched a groove along his skull. It split his helmet wide open, perfectly, as though with a knife. Then he passed out.

'We didn't contact Les, the rear gunner, until later. His oxygen tube had been severed, so, at 19,000 ft, he just passed out and never 'came to' until we got back over the Channel – he should have been dead. He woke up, by all accounts all 'googoowoowoo', with nobody to talk to.

'I yelled out to Bill, the bomb aimer, 'Have you jettisoned, yet?' He said 'No!' So I said, 'For Christ sake do so, we're on fire.' I could see down the fuselage: all the plates right down to the very end were aflame and the entire floor over the bomb load was crinkling red hot. Before you could say 'Jack Robinson' he'd called out 'Bombs gone'.[5]

About 220 night fighters were despatched, returning with claims for eleven Lancasters destroyed for nine own losses. In all, sixteen Lancasters failed to return. At Little Snoring there was no word from 'F for Freddie' skippered by John Anderson. The aircraft and crew were lost with no survivors.

'H-Harry' on 97 Squadron at Bourn flown by 21-year-old Pilot Officer Kenneth Painter crashed at Falkenberg with Sergeants' R. C. Saunders and T. Andrews the two survivors being taken prisoner. Sixteen Lancasters on 57 Squadron had left East Kirkby for Leipzig and one failed to return. 'E-Edward' skippered by Pilot Officer Frank Ernest Saville Parker was lost with no survivors. At Oakington two Lancasters on 7 Squadron were missing. 'H-Harry' piloted by 29-year-old Flying Officer James Westwood Leitch DFC RAAF had crashed east-north-east of Zwolle in Holland. Only Flight Sergeant F. W. Lashford the RAFVR flight engineer survived to be taken into captivity. A chartered accountant of Melbourne, Victoria, Leitch enlisted on 21 June 1941. He had completed 33 raids before he was killed and was awarded a posthumous DFC for his "determination to complete attacks no matter how hazardous or perilous the situation". He left a widow, Marion Helen Leitch, of Elizabeth Bay, New South Wales. 'A-Apple' skippered by 20-year-old Flight Sergeant Donald Moulton Watson of Willesden, Middlesex was shot down by Heinz-Wolfgang Schnaufer and crashed east-north-east of Assen in Holland with no survivors.

The crew of 'S-Sugar' on 103 Squadron at Elsham Wolds piloted by 30-year-old Warrant Officer Aubrey Huia Pargeter RNZAF, born Wanganui, the son of Thomas and Annie Sylvia Pargeter of Christchurch, Canterbury, New Zealand were on their 14th operation. They and Flight

---

5. *Ops: Victory at All Costs: Operations over Hitler's Reich with the Crews of Bomber Command 1939–1945, Their War – Their Words* by Andrew Simpson.

Sergeant Edgar Harold Willcocks RAFVR the 21-year-old second pilot from Saltash in Cornwall, who was flying as '2nd dickey', were lost without trace and are commemorated on the Runnymede Memorial. Nothing was heard from the crew on 'B-Baker' on 619 Squadron skippered by 22-year-old Pilot Officer Christopher Firth since take off from Woodhall Spa at 1718 hours. There were no survivors. Firth and his flight engineer, 24-year-old Sergeant George Myron Weighell left widows to grieve for their loss. Audrey Firth of Dunston, Gateshead County Durham, who was with child, would christen her daughter Christine after the father who never saw her.

At Kirmington two Lancasters were missing on 166 Squadron. All seven crew on 'K-King' flown by 21-year-old Sergeant Donald Bartley of Dulwich Village in London were killed. Flight Sergeant George Rodney Alexander Walkem the 26-year-old Canadian air bomber left a widow. On 'B-Baker' Sergeant Glyn Davies Hutchins the wireless operator was the only survivor on the crew skippered by 21-year-old Sergeant Stanley Tallintire Athey that was shot down near Köthen about thirty miles north of Leipzig after encountering very severe icing which caused a steep dive and total loss of control. Flight Sergeant Harry 'Hal' Millard the 33-year-old navigator left a widow, Florence Ruth Millard, of Balsall Heath, Birmingham. Margaret Ann Wardley, Sergeant Tom Kenneth Wardley the flight engineer's wife was also widowed with his passing.

Kelstern, high on the Lincolnshire Wolds northwest of Louth had despatched thirteen Lancasters. 'R-Robert' became the first 625 Squadron crew reported missing since its formation. All seven men on the crew of 20-year-old Canadian, Pilot Officer William Parmenas Cameron of Winnipeg, Manitoba died on the aircraft which crashed off Oosterend, Holland. Another 'R-Robert', one of 14 Lancasters on 405 'Vancouver' Squadron RCAF dispatched from Gransden Lodge, was flown by 27-year-old Pilot Officer Kemble Russell Wood RAAF of Sydney. On 17 December 1942 he had received a telegram saying that his wife Ethel Winifred Wood, of Strathfield, NSW had given birth to a daughter. It had taken Kemble six months to get out of staff work and get a posting to Path Finders. At Heavy Conversion Unit he picked up a 'headless crew' (one that had lost its pilot). This crew included three Canadians – Warrant Officer2 Ernest Charles Brunet of Montreal, Quebec, Flying Officer John Norman Ralston Redpath of Ottawa, Ontario and Warrant

Officer 1 O. O. Johnson the mid-upper gunner – which he thought might have been the reason that they had been posted to 405, the first-formed of the RCAF squadrons in Bomber Command and the only Canadian unit in the Path Finders. The crew, which was completed by RAF Sergeants' James Henry Lovelock, Francis William Bundy, 27, of Racedown, Bridport, Dorsetshire and Flight Sergeant William Henry Hedley, 22, whose wife May lived in Bury, Lancashire, had been together for about six months. Johnson said that they were all the very best of friends and that the life of a crew like theirs made the strongest friends that anyone could ever have.

An inveterate letter writer, Kemble Wood had penned his 99th letter to his wife Ethel on 16 October. She would not receive a hundred. Her husband and all his crew except Johnson were killed when 'R-Robert' crashed near Werlte at Harrenstätte. On 24 October funerals for those who died were held there. The next telegram that Ethel Wood received was the one telling her that her husband had been killed in action.[6] Johnson recounted later: 'At dusk we crossed the East Frisians doing a 15 degree weave, climbing. After crossing the fighter belt we stopped the weave and were climbing on track with 'George' in. The bomb aimer had just given the navigator a heading and distance on Bremen when we were hit by flak. I believe this was predicted because there was no lumpiness and no flak puffs to be seen. The port inner started on fire and the pilot said he would dive and put the fire out; at the same time he told the flight engineer to feather the engine. No answer from engineer, at once the starboard inner started to burn and pilot gave the order, stand by to bail out. At this time the rear gunner screamed for help. I left my turret and tried to get to rear gunners turret – but could not get through due to fire in the fuselage. Here my clothes caught fire. By this time we were about 15,000 ft. I returned to my turret and plugged in, but the intercom was dead. I left my turret and put on chute – at same time pulling air bottle on Mae West. I went back through fire to entrance door, and as I opened the door the fire from the starboard wing came through. I jumped through the flame and opened the chute when the fire on my clothes went out. I could not see the ground for cloud and fog and landed in a tall forest.'

---

6. See *Chased By The Sun: The Australians In Bomber Command in WWII* by Hank Nelson. (Allen & Unwin 2006)

Owing to suspected damage by flak to the fuel system, 'B-Baker' on 44 Squadron skippered by 23-year-old Pilot Officer Robert Hamilton Watts RAAF of Hampton, Victoria had to ditch at 0013 hours about sixty miles north east of Grimsby. The aircraft broke into two almost immediately after ditching and the pilot went down with the nose. Sergeant John Norman Connor the 22-year-old navigator of Glasgow dived off the wing presumably to help the pilot and neither was seen again. The remainder of the crew were rescued from their dinghy by the mine sweeper *Loch Moidart*. All were safe and they were returned to the Squadron. Pilot Officer Watts is buried in the Bergen-op-Zoom War Cemetery in Holland.

At Warboys no word had been received from two crews on 83 Squadron. 'E-Edward' piloted by 22-year old Warrant Officer Stanley Gordon William Hall of Harlescott, Shropshire is believed to have crashed in the sea. All seven crew are commemorated on the Runnymede Memorial. 'A-Apple' skippered by Squadron Leader Richard Manton was claimed by 23-year old Oberfeldwebel Heinz Vinke of 11./NJG1, flying a Bf 110G from Leeuwarden airfield for his 34th victory of the war and the first of his two kills this night.[7] There were no survivors. Harry de Waard and his friends looked for pieces of interest in the churned up, muddy meadow with pools of oil and burned wreckage and he found a charred purse which contained some coins and a medal. After he cleaned it, it appeared to have belonged to Flight Sergeant Roy Taylor. After the war, with the help of the International Red Cross, Roy's parents' Leonard David John and Augusta Miriam Beatrice Taylor of Greenford Green, Middlesex were traced. In 1946 they visited the grave of their only child in Assen. They thanked Harry de Waard for keeping their son's medal which was presented to his parents. Roy's father told Harry 'my wife and I will treasure it'.

---

7. Vinke's second victory was Lancaster ED555, one of 17 Lancasters on 100 Squadron that had taken off from Waltham (Grimsby) and flown by 21-year-old Pilot Officer Theodore Leonard Simpson, who was on his 25th operation. ED555 was shot down at 2225 hours and crashed at Eelderwald in Drenthe in the Netherlands with no survivors. Vinke had claimed his first aerial victory on the night of 27/28 February 1942. Following his 25th aerial victory, he was awarded the Knight's Cross of the Iron Cross on 19 September 1943. He and his crew were shot down and KIA on 26 February 1944, while on a search and rescue mission over the English Channel. Vinke was posthumously bestowed with the Knight's Cross of the Iron Cross with Oak Leaves.

On 20 October 1999 a stone monument to the Manton crew complete with a propeller blade and a commemorative plaque was unveiled in the crew's memory. It was constructed by a Dutch air war study group whose leader, Chris Timmer recruited Will Arents from Rolde near Assen to research the crash of JB154. Will's sub-group discovered that 86-year-old Mr. Dries owned the land on which the Lancaster crashed and during the 1970s had sold the area for sand excavations for motorway construction. During the excavation many aircraft parts were found but he remembered that a supervisor during the construction had found the propeller blade which he had taken home. On tracing the supervisor it was found that he still had the propeller blade in his garage. Mr. Dries and the group then decided that they should erect a memorial to the brave young men who gave their lives for Dutch freedom. Regrettably, in the summer of 2003 the propeller blade was stolen from the monument. On the advice of Squadron Leader Ed Bulpett, the Community Relations Officer at RAF Coltishall, Peter Rowland, husband of Albert's sister Ann and a former flight sergeant and 78th entry Halton apprentice, wrote to the O/C Battle of Britain Memorial Flight at RAF Coningsby telling the whole story and sending pictures of the movement. Peter Rowland had a call to say that the Battle of Britain Memorial Flight (BBMF) could help with a replacement propeller, which he duly brought back to his home in Lowestoft. It was packed up and sent to the Netherlands. In the meantime the Dutch Air Force had a damaged Stirling propeller blade which they gave to the Dutch air war group until they could arrange for an anti-theft device to be fixed to the Lancaster propeller blade donated to reinstate the monument. The replacement Lancaster propeller was not installed in time for the 60th anniversary of JB154's last flight but was ready in time for Dutch Liberation Day, 5 May 2004. The custom in the Netherlands is to hold a two minutes' silence at 8pm on the evening of 4 May. As in all previous years since 1947, school children from Assen had placed flowers on the seven graves during the afternoon of 4 May. Later that evening, Will Arents and his group held their two minutes' silence, ended with a reading in English of Rupert Brooke's poem, *The Soldier. If I should die, think only this of me; that there's some corner of a foreign field that is forever England.*

On Liberation Day itself several hundred people made their way by car, bicycle and on foot to the fields outside Assen where the plane had crashed in 1943. The Mercurious Marching Band played military music and a lone piper played the bagpipes. Speaking in English and Dutch, Will Arents welcomed everyone and reminded everyone of the part played by the RAF in liberating the Netherlands in 1944. The applause as Ann Rowland unveiled the memorial turned to cheers as the RAF's last remaining Lancaster appeared over the trees. In all, it performed three low passes, flying as low as 300 ft – the roar of the four Rolls Royce Merlin engines drowning a mixture of cheers and sobs. Several of the crowd became very emotional – many of the older men and women had stories of German occupation they wanted to tell. Some were openly crying. One man said that the Netherlands was liberated 'by luck and the Lancasters'. Ann Rowland herself showed mixed emotions – sadness at the thought of her brother's death and happiness at the way in which this little Dutch town had adopted the seven airmen almost as their own. Will Arents was highly delighted at the size of the crowd at the memorial's unveiling. 'We are astonished that so many people have shown their interest,' he said. 'But they were all truly shocked at the damage done to the original memorial.'

By now, the Battle of Britain Flight's Lancaster had landed at the nearby Groningen Airport and again, hundreds of onlookers had parked at every vantage point. Those lucky enough to have an invitation to view the Lancaster were allowed much closer than they ever felt possible. Will Arents, Ann Rowlands and others were invited into the cockpit whilst many others were shown around the aircraft.

Flight Lieutenant Jack Hawkins of the BBMF was well-used to this sort of emotional reception for his historic Lancaster. 'We always get a fantastic welcome when we come to the Netherlands,' he said. 'Usually we perform a fly-past and then just head off back to UK. But today, it has been very special – we knew that we were invited to land the aircraft and meet many of the people involved in the JB154 story. I've brought the ground crew just so that they can see what an emotional thing this is for the people of Assen'. The crew were photographed and filmed to within an inch of their lives and mementoes were presented to all of them, including the ground crew.

To a huge cheer and lots of waving handkerchiefs, the Lancaster took off for its flight back to Coningsby … .and Ann Rowland went back to pay just one more quiet visit to her brother's grave.

Chapter 6

The Valley Of Death

Half a league, half a league,
Half a league onward,
All in the valley of Death
Rode the six hundred.

The Charge Of The Light Brigade
by Alfred, Lord Tennyson.

On the night of 15/16 March 1944, 617 Lancasters (and 230 Halifaxes and 16 Mosquitoes) took part in the raid on Stuttgart.

Spread across a variety of hills and deep valleys (especially around the Neckar River and the Stuttgart basin), the city of the same name, known as the "German Coventry" was an important rail hub and a centre of industry, home to the Bosch, Daimler-Benz and the SKF ball bearings factories. Unsurprisingly, Stuttgart had become a constant target for RAF Bomber Command. Shelters dug into the sides of the surrounding hills saved many lives. By 1944 the city was defended by eleven 88mm and thirty-eight 20mm to 40mm anti-aircraft gun batteries. There was also a Luftwaffe fighter base south of the city at Echterdingen. The landmark brick observation tower which had been formally opened on 19 September 1891 in Burgholzhof was used by anti-aircraft spotters during raids and the thirty metre high Pragstattel Flakturm (Flak Tower) stood just north of central Stuttgart while three cone-shaped Winkeltürme, which could each hold 500 people, were built around the city.

On 11 March 1943, 314 RAF bombers arrived over Stuttgart. Pathfinder units claimed to have illuminated targets in the city, but most of the bombs dropped that night fell in open country and on dummy Pathfinder indicators, which the Germans were using for the first time. A total of 112 people died and 386 were injured when Vaihingen and Kaltental were hit, resulting in the destruction of 118 houses. Six Halifaxes, three Stirlings and two Lancasters (3.5% of the total force), were lost on the operation. More raids, by RAF Bomber Command and the US Eighth Air Force followed and on 3 November 1943, Arthur Harris listed Stuttgart among nineteen cities he claimed had been 'seriously damaged' in a report of Bomber Command's activities to Prime Minister Winston Churchill.

In 1944 Stuttgart was heavily bombed by the RAF. On the night of 20/21 February when 598 bombers visited the city only seven Lancasters and a single Halifax were lost to enemy action. Over the next two nights, a total of 27 Mosquitoes attacked Stuttgart. On 25 February, in the final mission of the 'Big Week', fifty Flying Fortresses of the 1st Bombardment Division, 8th Air Force added their weight of bombs to the city. On St. David's Day, 1st March 557 Lancasters, Halifaxes and Mosquitoes were largely shielded from attack from German fighters by thick cloud on routes to and from Stuttgart. At Kirmington 28-year-old Flying Officer William E. 'Jonah' Jones of the Intelligence Operations staff had conducted the briefing for fifteen Lancaster crews on 166 Squadron and everything had gone smoothly. 'According to the crew reports it was a very good attack and with a very low loss rate – only three Lancasters and one Halifax missing, or less than one per cent."[1] The hiatus that followed saw raids continue only to targets in France and mainly they were carried out by the Halifax squadrons while the Lancaster groups drew breath, repaired battle-damaged machines, awaited new ones and welcomed fresh crews to fly them. Finally, and with the moon on the wane, over 600 Lancasters were available for the next round of operations when on the night of Thursday 15/Friday 16 March, Stuttgart was on the Battle Order for the third time in a little over three weeks. The unfortunate city would be the destination for a force of 617 Lancasters, 230 Halifaxes and 16 Mosquitoes that was assembled in a hive of activity at every station in Bomber Command. At Kirmington 166 Squadron was able to report a record number of 22 Lancasters ready for action. At Wyton, 83 Squadron briefed sixteen crews, with Wing Commander Joe Northrop the most senior pilot, backed up by the newly-promoted Squadron Leader Ernest Neville Monkhouse Sparks.[2] Take-off at Wyton took place between 1932 and 2005 hours with Joe Northrop getting away a little later than planned.

---

1. *Bomber Intelligence* by W. E. Jones (Midland Counties Publications, 1983).
2. Sparks would be shot down twice in 1944, first on Mailly le Camp on the night of 3/4 May when he evaded and on 27/28 August when he was shot down on Königsberg and taken prisoner. He was awarded the DFC on 25 January 1946.

At Dunholme Lodge airfield snuggling between the parishes of Welton and Dunholme in Lincolnshire where the runways went up a hill and down the other side, Flight Lieutenant Terence Hugh Fynn, a 'second tourist' on 44 (Rhodesia) Squadron took 'M-Mother off at 1931 hours and headed for Selsey Bill just to the west of Bognor Regis. Then he headed south and out to sea before crossing to the Le Mans area in France, then Switzerland and finally he headed north to Stuttgart. Flying Officer John Raynor Berrington, a 20-year-old pilot who had just joined 44 (Rhodesia) Squadron had asked if he could fly his '2nd dickey' flight with Fynn, who was a fellow Old Georgian at St. George's College in Salisbury, Rhodesia and a former classmate of Fynn's youngest brother John. Terry Fynn had been born in 1916 in King Williams Town in the Eastern Cape, South Africa. He left school in 1933 and went to the UK to become a Jesuit priest. He did six years training, but eventually decided that it was not for him. By then the war had broken out and, after having some doubt, he joined the RAF. He qualified as a pilot and after his first tour of operations he spent a year in Canada training pilots there. Fynn's crew – none of whom were Rhodesian – said that it was a bad omen to have two Rhodesians in the crew and the skipper should turn the novice pilot down. Pilot Officer Mervyn Adder was 21-years-old and came from Hull where he had been a clerk in county court. Sergeant Gerard Clark the 22-year-old air bomber from Harborne, Birmingham had been Chief Instructor (Building) at the Government Training Centre. Sergeant William Cameron Jack the 23-year-old Scottish air gunner was from Barrmill, Ayrshire; the son of a railway checker. Sergeant Ronald Johnson the 23-year-old wireless operator, the only married man in the crew who was from York had worked for a confectionary company. He loved doing the "boogie woogie" and was a terrific swing dancer. Sergeant Dennis David Orme the 20-year-old flight engineer came from Marylebone, London. Sergeant Joseph Sagar the 19-year-old air gunner was from Bramhall, Cheshire. Fynn told them not to be superstitious and accepted Berrington as second pilot.[3]

At Bardney one of the Lancasters chalked on the 9 Squadron blackboard was 'H-Harry'. In the pilot column was the name of Squadron

---

3. See *The Crew of Lancaster ND576 KM-M 44 Sqn, 15th March 1944* written and researched by Dorothy V. Ramser (May 2020).

Leader Raymond Backwell-Smith. He had served in India for a time and originally trained as a fighter pilot, which stood him in good stead because he would often 'fling the Lancaster all over the sky to try and shake off an attacker. The previous 29/30 December, Backwell-Smith's crew had got Lancaster 'D-Donald' back from Berlin on three engines after the outer starboard engine had been set on fire following an attack by a night-fighter. Sergeant Norman Sirman his flight engineer turned off the fuel supply to the engine and activated the extinguisher while his pilot feathered the propeller. Airspeed was reduced to 160 mph but with a strong headwind the Lancaster was only making 120 mph. Backwell-Smith made a wide circuit of the runway to the left at the emergency landing airfield at Woodbridge to keep the two good engines working before reducing speed from 140 mph and then 90 mph and Sirman closed the throttles and Backwell-Smith applied the brakes. Sirman then shut down the outer engines, leaving the single inner engine for taxiing.

At takeoff time for Stuttgart the crew assembled at the rear of their Lancaster. Thirty minutes' later Backwell-Smith started the two inner engines using the starter battery motor and booster. Once fired, the two outer engines started up. Engine revs 1,500 rpm, raised to 3,000 rpm, brakes released, taxi speed increased to 100 mph, backward pressure on the control column as speed increased to 125 mph to prevent stalling and at 500 ft the brakes were applied to stop the wheels spinning. Undercarriage retracted, the flaps raised, strict radio silence maintained and at 22,000 ft the airspeed increased to 200 mph. Their estimated time of arrival at the target area was just before midnight.

In an attempt to avoid contact with night fighters the force was split into two parts and the bombers flew a roundabout route over France nearly as far as the Swiss frontier before turning north-east to approach Stuttgart. This deception worked well until just before the bombers reached the target when 93 aircraft of 1st Jagdkorps were fed into the bomber stream. On 'H-Harry' the flight was uneventful until the crew neared Stuttgart, when they were attacked by a night fighter. 'We immediately took evasive action' wrote Norman Sirman 'but despite this the rear of the aircraft was very badly damaged and the three crew members in the back of the aircraft were killed during the attack. I was wounded in my right thigh. The second attack caused more damage to the aircraft and the third attack caused complete loss of all controlling mechanisms.'

Their attacker was Oberleutnant Heinz Rökker, Staffelkapitän, I./NJG2 who was flying a Ju 88. Rökker had served in I./NJG2 for almost two years and had destroyed his first 'Viermot' on 23/24 August 1943 when he shot down a Lancaster twenty kilometres southwest of Berlin for his seventh victory. Recently he had reached double figures in the victory tables. Now, on the night of 15/16 March, when his flight commenced at Langensalza at 2042 hours and terminated at 0047 hours at Frankfurt am Main, he would claim a hat-trick of victories beginning with 'H-Harry', which he picked up 50 kilometres west of Strasbourg at 2226 hours. Rökker later wrote: 'I had problems with this attack. For this reason, I remembered it well, whereas with the high number of attacks that I made, I often remembered little of the individual details. It was a protracted air fight of a type, which I only experienced twice and that is why I remember it. At the beginning of the attack we had to attack from behind and below using our front guns, but a short while later we were able to attack the fuselage with our rotating gun. After the first attack with the front guns, I had, indeed, observed a hit, but the aircraft did not catch fire. The pilot pushed the plane into a nose dive and in doing so made a successful defensive tactic, as English pilots had been trained to do. He flew defensively in climbing and turning sharply to the left and losing height, the aptly named 'corkscrew'. In the pursuit of him, I hit him again but could not see any flames. The rear gunner [21-year-old Flight Sergeant Eric Alfred Birrell from Mount Gambier, South Australia] did not shoot back. Either he had been killed in the first attack or he could no longer turn the heavy weapon machinery. The mid-upper gunner [19-year-old Sergeant Brian Glover from Didsbury, Manchester] gave defensive fire but it went wide. Finally, I lost sight of the enemy.' Birrel, Glover and Flight Sergeant Ronald West the 21-year-old wireless operator from Nottingham were killed.

'According to our rules of engagement' continues Rökker, we identified enemy aircraft only as a 'Viermot' ('4-engined bomber') because for me and the crew it was very difficult to differentiate between the Lancaster and the Halifax at night, as they both had four engines and a double tail unit. It was the first and last time that I had an air fight with an enemy, which I was unsuccessful in visualising and following to the end (no fire observed). Therefore, the shooting down was only provisionally awarded to me.'

'Backwell-Smith gave the order for immediate evacuation of the aircraft' recounted Norman Sirman. 'Following this all communication was lost. The bomb aimer was the first to jump followed by the navigator. After disabling the radar equipment, including 'Monica', I gave a parachute to the pilot and jumped out followed by him.'

Flying Officer Herbert Keith Sheasby, the 31-year-old navigator and Pilot Officer Douglas Raymond Eley the 20-year-old Canadian bomb aimer from Victoria, British Columbia did not survive their parachute jumps. Sheasby left a widow, Marjorie Irene of Shirley, Warwickshire. Backwell-Smith's parachute got caught in an electricity pylon and burned and he dropped to the ground, breaking several bones in his back. Sirman had a cannon fire wound in the leg and as he rolled through the escape hatch, the aircraft lurched and he gashed his head on the metal.

'As I came down I realised that the sky was bright although it was nearly midnight. I supposed it was the fire from the bombing in Stuttgart. When I reached the ground, I gathered my parachute together with the intention of hiding it ready to escape to the south and then found that I couldn't stand up due to the injury in my leg. I fainted. The next thing I remember was a soldier prodding me and speaking to me. A large number of people had gathered and two civilians helped me up. One of them explained in very good English that he was the local schoolmaster. They helped me to walk to a cottage nearby. A woman (whom I was to know was Frau Kaefer) was very kind and helped me to dress my wound and gave me a drink, which I discovered was Schnapps although I had thought it was water and drank it rather fast! She had a very young boy with curly hair and I gave him chocolate, which I carried in my pocket. He enjoyed this very much.

'The whole atmosphere in the house changed when a large man in a long, black mackintosh entered. He ordered everyone out of the room except the soldiers. Immediately after this Raymond Backwell-Smith was dragged in and it was clear that he had a serious back injury as his legs were not working. The Gestapo man kicked him in the back and when he went to do it again I thrust my leg in his way, thwarting him. He did not like this. Later I was taken to a place where a French doctor, who was a prisoner of war, removed the bullet from my leg. The Gestapo man refused to let him use any local anaesthetic. I was then put in a cell and sometime later Raymond was brought in encased in a plaster cast from neck to thighs. He had been told his back was broken but fortunately the bones had not moved. We were looked after by French prisoners of war with an armed German guard outside the door.

'Several days later I was taken away by the man in the black mackintosh. I was put in an underground cage below a large building and the treatment I received is still the source of regular nightmares. I do not know how many days I was there but was eventually rescued by a Luftwaffe officer after he had had a long and noisy argument with the people holding me. He spoke good English and was kind to me, taking me to his airbase, where I was given medical and personal care and, when I was well enough, allowed to mix with the airmen for meals, having given my parole that I would not try to escape. At that time I was not in a good enough condition to escape anyway. Eventually I was sent to Dulag Luft for interrogation and onwards to Stalag Luft I prisoner of war camp in Barth, Pomerania.'

The opening of the raid was delayed by adverse winds and the marking affected by ten-tenths cloud, into which the TIs (Target Indicators) disappeared. The skymarkers were somewhat sparse and seemed to form two clusters, one to the north-east and one to the south-west and it was at these locations that the attack developed. The 83 Squadron crews bombed from 17,000 to 19,800 ft and all would return safely although Joe Northrop remembered the raid well for he was lucky to get back from it.

"Planning for the attack was based on a Met forecast that a frontal system would clear the target area by our time on target leaving clear skies for a visual marking to be carried out. I carried an additional crew member for the trip, our most experienced visual bomb aimer (Warrant Officer Gray). His task was to identify the aiming point by the light of 4.5 flares dropped by an Illuminator aircraft just ahead of us. He would then mark it accurately with the huge green TIs we carried in addition to dropping our bomb load. Other Pathfinder aircraft would then back up our initial markers with further greens thus providing an unmistakeable aiming point for the main force of bombers as it approached the target. All was going well as we approached the target after an uneventful trip above heavy clouds; then the cloud started to break up as forecast and we gradually lost height to 8,000 ft and positioned the aircraft in readiness for the final run up, carefully checking our position and timing. Up to that moment there was no reaction from the ground defences ahead although we had run into the downwash of an aircraft several times and assumed it must be our Illuminator. Usually, such a reaction either meant that AI fighters were operating in the vicinity or the AA guns and searchlights were tracking us and lying 'doggo' until they were sure of a good fix. Now on the last stages of the run up I opened the bomb doors and listened to the set operator doing the countdown to the time of target illumination. With only seconds to go the aircraft ahead of me was suddenly 'coned' by searchlights and the heavy flak below opened up into the cone. They must have scored a direct hit with the first shells for the aircraft disintegrated in a vast sheet of flame lighting the whole area momentarily then dying away as a thousand pieces of burning debris appeared to float down in our path, colliding with our aircraft in a series of bangs and crashes.

"The smooth synchronised notes of the four Merlins changed as the engines vibrated badly in their mountings indicating airscrew damage. The aircraft had a decidedly port wing down attitude and an icy gale was blowing through the cockpit from breaks in the front turret and damaged perspex on the pilots clear vision panel. To make matters worse the searchlights then held on to us and shells were bursting all around as the guns gave us their undivided attention. I called to Gray the bomb aimer to hang on to the TIs but to let the bombs go and felt the aircraft lurch as they went and he closed the bomb doors. For the next few minutes I twisted and turned and lost height as I tried to shake off the searchlights. Suddenly we were in darkness as they looked elsewhere for another victim. I asked the navigator for a direct course for home and checked the crew to see if they were all OK; meanwhile the engineer juggled with the throttles and pitch controls and managed to cut out some of the vibration.

"We were now at around 6,000 ft and fast approaching the solid wall of cloud over which we had flown on the way in. I got Forster, the Australian flight engineer, to use the Aldis lamp to check for serious damage to the engines and wings before we entered the clouds. He reported damage to the leading edge of the starboard wing and a large piece of Lancaster fuselage jammed into the port wing causing the drag and the left wing down attitude. Happily there seemed to be no damage to the engines so the vibration must have been caused by bent or damaged airscrew blades. After a nightmare journey flying through thick clouds in freezing conditions we eventually broke cloud over the North Sea and limped into Wyton nearly seven hours after takeoff."⁴

First Jagdkorps returned with claims for thirty kills for the loss of nine aircraft and crews. Bomber Command lost 34 aircraft (ten were Halifaxes) shot down and two Lancasters that were badly damaged by enemy fighters over France headed for Switzerland. An hour before reaching Stuttgart, Lancaster 'F for Freddie' on 57 Squadron that had taken off from East Kirkby at 1857 hours had all four engines set on fire. Pilot Officer Samuel Cunningham Atcheson the 24-year-old pilot, who was from Drumquin, County Tyrone, ordered all his crew to bail out once they had crossed into Switzerland but only one of the gunners, Flight Sergeant Kenneth A. Reece, made it before the Lancaster crashed at Saignelegier. "Sam", as Atcheson was more fondly known, had been recommended for the DFC on 5 March after his 19th operation, including twelve raids on Berlin. His award was made posthumously on 24 March for his devotion to duty while serving in Bomber Command.

Flight Lieutenant Walter 'Bill' Blott RAFVR on 15 Squadron who had taken his Lancaster off from Mildenhall at 1915 hours was attacked outward-bound and east of Vesoul in the Haute-Saône département of France at 2211 hours after crossing the Rhine north of Basle. His attacker was Hauptmann Eckart-Wilhelm von Bonin of Stab II./NJG1 flying a Bf 110G-4 from Sainte-Dizier airfield in Haute-Marne. At the last moment Sergeant G. R. Mattock the 22-year-old married flight engineer, a former welder from Winchester, who was standing watch in the astrodome, saw the Messerschmitt that had slipped in from behind and below unnoticed by Sergeant D. Murphy the rear gunner in a classic Schräge Musik attack and was sweeping up in a left turn. The upward pointing cannon holed the port-wing tank, disabled the left inner Merlin engine, made the intercom unserviceable and rendered the blind-flying panel useless. Glass shards, oil, smoke and petrol burst into the cockpit and Blott felt a severe blow on his left elbow; a metal fragment had hit him, rendering the arm useless. The Lancaster flew on but soon started shaking uncontrollably and Blott gave the order for the crew to bail out. They safely vacated the aircraft over the Lac de Neuchâtel in Switzerland

---

4. Cited in *Joe: The Autobiography of a Trenchard Brat* by Wing Commander Joe Northrop DSO DFC AFC PFF.

before it crashed among some fruit trees 50 metres from a farmhouse at Golaten west-northwest of Berne. One of the crew had suffered a broken leg, another a bullet wound. Blott landed heavily on his back in a wood near Kallnach, halfway between Berne and Bielersee. Making contact with a Swiss woman, he was given a bed in an inn nearby for the night. After breakfast the following morning he was visited by Swiss military police officers, who told him that they would be returning later in the day after they had collected others of his crew.

After three weeks at the Gurten-Kulm hotel the crew was transferred to the internment camp at Adelboden and in May were repatriated. They travelled by train to Irun in Spain and on to Gibraltar on 19 May before being flown home five days later.

On the way back at 2241 hours Lancaster 'E-Edward' on 514 Squadron which was on its 13th operational flight and was piloted by 21-year-old Flying Officer Kaiho 'Tommy' Penkuri RCAF of Port Arthur, Ontario, was shot down at 2241 hours although there is some doubt as to who delivered the killer blow. Hauptmann Eckart-Wilhelm von Bonin claimed a '4-mot' at St. Orthelemy northeast of Lure for his second victory of the night. Oberfeldwebel Kellerman of L.Beobachter.St.7 claimed a Lancaster 'probable' 10 km west of the Überlinger See, the lower part of Lake Constance, at 2258 hours as his 14th victory. 'E-Edward' exploded and crashed between Blondefontaine and Villars-le-Pautel in the Haute-Saône with no survivors. The 8-man crew included Flight Sergeant Kenneth Drummond RNZAF the 23-year-old 2nd pilot, who was on his first op on the squadron, flying his 'second dickie' trip.

Thirteen Lancasters on 408 'Goose' Squadron RCAF had begun taking off from Linton-on-Ouse starting at 1901 hours. All seven crew on 'P-Peter' piloted by 23-year-old Flight Sergeant Norman Andrew Lumgair RCAF of Thornhill, Manitoba were killed when their Lancaster was shot down by a German night fighter and crashed at Hilsenheim 9 km east of Selestat in France. Oldest man on the crew was 30-year year-old William Lawrence Doran, the Canadian wireless operator who left a widow, Marjorie Evelyn of Edmonton, Alberta. Sergeant Douglas Cruickshank the 23-year-old flight engineer was a Yorkshireman from Harrogate. Pilot Officer George Parker the 29-year-old Canadian navigator from Morinville, Alberta left a widow, Patricia Anne Parker. Pilot Officer William Taylor the 23-year-old air bomber was from

Nottingham, Saskatchewan. Both air gunners, Sergeants' Robert Henry Hudson RAFVR from Sileby, Leicestershire and Robert George Alfred 'Bud' Burt RCAF of Brampton, Ontario were just nineteen. 'E-Easy' piloted by 28-year-old Flying Officer Alexander Colborne Colville RCAF, of Bowmanville, Ontario had been the third Lancaster to takeoff, at 1903 hours but nothing further was heard and he and his crew were listed as lost without trace. Colville enlisted in Toronto on 9 September 1940. Graduating as a flight sergeant air gunner, he later re-mustered as a pilot and received a commission to pilot officer. Due to bad weather there was little flying activity from 25 February until 7 March when he flew as '2nd dickey' on his first operation. Sergeant Michael Yorke Zisslin Kalms RAFVR the 19-year old flight engineer, born in July 1924 in West Ham to Jack and Russian born Debora (née Zisslin) had been living in Leytonstone. The family, who were of the Jewish faith, had moved to Edgware in Middlesex after the family home was destroyed by enemy bombing. A motorcycle enthusiast and keen boxer, after leaving school, Michael had become an engineering apprentice at de Havillands before joining the RAF on 8 October 1941 aged seventeen. Flying Officer Moody Albert Siddons the 30-year-old Canadian bomb aimer and the oldest member of the crew came from Langford, British Columbia. Warrant Officer1 Arthur Coles Kitchener Hodson RCAF the 27-year-old WOp/AG was from New Westminster, British Columbia. Sergeant Dennis Vivian Davies RAFVR the 21-year-old rear gunner was from Rhondda, Glamorgan. The two others who died were Flying Officer William Van Fossen Reid RCAF the 21-year-old navigator of Unity, Saskatchewan and Sergeant Francis Ernest Albert Smith RAFVR, mid-upper gunner.[5]

Two of Colville's Bowmanville High School friends were killed on Stuttgart. Flying Officer Kenneth Arthur Cole a 20-year old air gunner on 405 City of Vancouver Squadron at Gransden Lodge died when 'H-Harry' was claimed shot down over the target area by Unteroffizier Emil Nonnenmacher of 8./NJG2. Cole's skipper, 23-year old Flight

---

5. Colville's 24-year-old brother William Freeborne Colville of the RCAF East Coast Squadron had been killed on 6 May 1942 at Torbay, Newfoundland and Labrador. Flying Officer John 'Sandy' Spencer Colville, the third brother was shot down on 18 August 1944 while flying a Typhoon on a reconnaissance flight in the area of Orbec, France and he apparently bailed out but was too low and his parachute failed to open. They are the only known three pilot brothers killed while on Commonwealth active service. Mrs Alex J. Colville proudly wore three silver crosses in their memory.

Lieutenant Allan Blake Fyfe and the wireless operator and the flight engineer also died. Three crew members survived and were taken prisoner. Pilot Officer Lorne Edgar Yeo the 21-year-old rear gunner on 'P-Peter' on 426 'Thunderbird' Squadron at Linton-on-Ouse died with the rest of 23-year-old Pilot Officer Arthur Gerald Sylvain Simard's crew after being attacked by Unteroffizier Bruno Rupp of 4./NJG3 southwest of Stuttgart at 2328 hours and crashing at Boblingen Forest. Simard, who was born in Tilbury East, Ontario, had enlisted in the RCAF on 21 February. His brother Reginald was killed in action on 31 August 1944 while serving in the Perth Regiment.

Earlier, at 2310 hours, flak had claimed 'U-Uncle' on the 'Thunderbird' Squadron 10 km southwest of Stuttgart at 19,865 ft. Warrant Officer2 C. H. McIlwain RCAF and his six crewmembers abandoned the aircraft before it crashed in the Böblingen area following total engine failure and they were taken prisoner. Flight Lieutenant William Frederick Nicholls RCAF the 23-year-old bomb aimer of Winnipeg, Manitoba died from tuberculosis and inflammation of the brain on 13 January 1945 whilst at Stalag Luft IV. He was buried at the PoW camp at Gross Tychow. Post war his grave could not be located and he is commemorated on the Runnymede Memorial.

Flight Sergeant Frank Grafton Hodgkins on 625 Squadron had taken DV194 off from Kelstern at 1908 hours. Outward-bound, the aircraft was claimed shot down by Feldwebel Alfred Rauer of 3./NJG1 who was flying a Bf 110G-4 from Laon-Athies airfield in France and it exploded scattering debris near the village of Obereschach. Hodgkins, who after leaving Worcester Royal Grammar School in 1938 and joined the RAF as an engineering apprentice had volunteered for aircrew early in the war and after training in Canada he qualified as a pilot. All seven crew including Sergeant Harold Edward Chapman the 19 year-old rear gunner of East Greenwich, London and the wireless operator, Sergeant Vlieger Alexander Joseph Smith of the Royal Netherlands Naval Air Service were killed. Sergeant Allen Edward Roy Ward the 28-year-old air bomber left a widow, Margaret of Blackpool, Lancashire. 'J-Jig' skippered by 21-year-old Pilot Officer Derrick John Gigger of Eynsford, Kent that took off from Kelstern at 1855 hours disappeared with all eight crew and was presumed lost in the sea off the coast of France. Gigger, who had been born in France in 1923, is buried in Dannes Communal Cemetery, France.

At North Killingholme 550 Squadron dispatched 16 Lancasters – one failing to takeoff owing to un-serviceability – and two were lost in action. Major Rolf Leuchs of Stab II./NJG6 flying a Bf 110G-4 from Echterdingen airfield is believed to have shot down 'X-X-ray' piloted by 31-year-old Flight Lieutenant James Fraser Craig DFC RNZAF 25 km southwest of Sainte-Dié, France at 2235 hours. The bulk of Craig's crew was previously a 'headless' crew after the loss of their pilot on his '2nd dickey' flight on 19/20 February. Craig and four of his crew were killed and Sergeant H. C. Petty the mid-upper gunner was taken prisoner. Unteroffizier Herbert Koch of 6./NJG1 who was flying a Bf 110G-4 from Sainte-Dizier airfield in Haute-Marne claimed three Lancasters. Outward-bound, 'Q-Queenie' on 101 Squadron, which monitored and jammed German night fighters using Lichtenstein radar sets and was flown by Flying Officer James Clegg GM RAAF, crashed at Hilsenhein, 22 miles southwest of Strasbourg. All eight crew including Sergeant Joseph Bagnall Bull RAFVR the Specialist ABC operator were killed. 'J-Jig' on 550 Squadron at North Killingholme flown by 24-year-old Flight Lieutenant Jacque Simon Gustave Crawford of Harrow, Middlesex crashed southeast of Schlettstadt, France. All seven crew were killed. Crawford left a widow, Pauline May Crawford. Koch's third claim was A4-K2, one of 22 Lancasters on 115 Squadron at Witchford skippered by 21-year-old Pilot Officer James Menzies Rodger RAFVR of Newmilns, Ayrshire that crashed at Geitendorf, in the north east district of Stuttgart following a night fighter attack at 2340 hours, shortly after bombing. Sergeants' Reg Favager, wireless operator and Lawrence Joseph Casey rear gunner were the only survivors. Their five fallen crewmembers are buried in a mass grave at the eastern end of Seckselberg Cemetery. Casey had enlisted on 6 February 1939 and volunteered for aircrew duties. Following training in Iceland on 21 July 1941 he served on Whitleys on 58 Squadron and Stirlings on 623 Squadron and also saw service in Boulton Paul Defiants at Linton on Ouse before taking a 9-month posting as a gunnery instructor at 8 AGS at Evanton in Scotland. Lawrence married Sarah Alice Twiss, in Edinburgh and they had three children.[6]

---

6. Lawrence Casey died in 1952. Oberleutnant Herbert Koch scored his 23rd victory (a Lancaster) on 24/25 April 1945 which has gone down in history as the Nachtjagd's 7,308th and final victory of WWII.

'L-Leather' piloted by Flying Officer Arthur Ganderton was one of twelve Lancasters on 35 Squadron at Graveley detailed to attack Stuttgart. Nothing was heard after takeoff. The Lancaster had crashed at 2300 hours at Niedersachsen near Lahr in the Black Forest area of Germany after being attacked by 27-year-old Oberleutnant Günther Köberich of 6./NJG2, flying a Ju 88C-6 from Quakenbrück airfield. Ganderton and his flight engineer, Sergeant John Stewart Martin, Sergeant Harry Roy Lowman, gunner and Flight Sergeant Colin Oswald Gibbons DFM, bomb aimer, bailed out and were captured. Sergeants' George Bagnall the wireless operator; Joseph William Samuel Burden the 30-year-old navigator and Arthur Henry Weller the 41-year-old air gunner were later buried at Lahr Cemetery. Accordingly, on 16 March the squadron regretfully informed Helen Weller of Brighton and the other crewmembers' next of kin that they were 'missing as a result of air operations'.

At 2312 hours 'F-Freddie' on 7 Squadron, which 27-year-old Pilot Officer Douglas Arthur Carter RAAF had taken off from Oakington at 1939 hours, was shot down at 2312 hours southwest of Stuttgart by 23-year-old Hauptmann Gerhard Ferdinand Otto Raht of 4./NJG3 for his 22nd victory. 'Freddie' was seen approaching the German village of Tannheim northwest of Donaueschingen (Baden-Württemberg). One witness said that it was burning fiercely in the air and was seen to crash and exploded on impact, leaving a crater 100 ft wide and 25 ft deep. According to Fraulein Marka Schwaab of the Burgomesiter's officer, wreckage was spread over a very wide area. Other witnesses stated that pieces of the crew were found in the surrounding trees. The bodies of the eight crew members were recovered by a Luftwaffe unit and were buried on 21 March in the village cemetery at Villingen. Carter left a widow, Thora Cecilia Carter, of Bournemouth, Hampshire. Flying Officer Kenneth Charles Dyer RAAF the 26-year-old navigator had been married to Doreen Yvonne Dyer, of Epping, New South Wales. The wife of Pilot Officer Peter Hanson Hamby RAFVR the 2nd pilot of Sheffield was also widowed.

At 2315 hours Günther Köberich downed his second Lancaster. The 49 Squadron crew skippered by 23-year-old Pilot Officer Thomas William Waugh who were on their first operation, had been the third aircraft to take off from Fiskerton, at 1905 hours. Having dropped their bomb load on Stuttgart, Waugh realized that they wouldn't make the journey home so he turned south and headed towards Switzerland. Just 18 minutes flying time from the Swiss border the Lancaster was shot down at 20,000 ft above Bolstern southwest of Saulgau in Baden-Württemberg. There were no survivors. Sergeant John Grenfell Wise the rear gunner was 19-years-old. The crew are buried in a collective grave at the Durnbach War Cemetery.[7]

At 2319 hours Oberleutnant Fritz Lau of 4./NJG1 flying a Bf 110G-4 from Ste. Dizier airfield shot down 'C-Charlie' on 12 Squadron at Hartheim am Rhein, Baden-Württemberg which 23-year-old Warrant Officer Donald Russell Knowles had taken off from Wickenby at 1840 hours. The engines were set on fire and the bomb load exploded killing six of the crew. Only Flight Sergeant Harold Edgley and Sergeant H. J. Sishton survived. The crew included an eighth crewmember; 25-year-old Flight Sergeant Albert Ernest Hammond RCAF, who was on his '2nd dickey' flight. For his third victory of the night, at 2345 hours, Günther Köberich claimed 'B-Bertie' on 97 Squadron piloted by 34-year-old Flight Lieutenant William Alexander Meyer DFC of Piccadilly in London, who had taken off from Bourn at 1920 hours. 'B-Bertie crashed at Zillhausen 5 km ESE of Balingen killing Mayer, Pilot Officers' Reginald Charles Pike DFM, wireless operator and James Campbell McLeish DFC the WOp/AG and air gunner, Archibald MacArthur Barrowman DFC of Toronto, Ontario. Flight Sergeants' Thomas Roy Shaw DFM the flight engineer and Albert Edward Roberts DFM, air gunner and the navigator, Flight Lieutenant Bernard John Starie DFC died too.

Two pilots on 460 Squadron RAAF at Breighton which dispatched 24 Lancasters on the raid described the attack on Stuttgart as "irresponsible, scattered and unimpressive" and criticised the "wild bombing" which put most of the weight of the attack near fourteen villages to the south-west of the target. The Australian squadron lost one Lancaster missing in action. No trace of 28-year-old Pilot Officer George Edward Parkinson's crew on 'E-Easy' could be found and they are commemorated on the Runnymede Memorial. Parkinson left a widow, Violet Kathleen Winifred of Parramatta, New South Wales. Born in Blackburn in

---

7. Köberich was killed in an American air raid on the airfield at Quakenbrück on 8 April 1944.

England, Parkinson had enlisted in the RAAF at Sydney, NSW on 17 August 1941.

Seventeen Lancasters had been dispatched by 463 Squadron RAAF and no word had been received from the crew on 'J-Jig' piloted by 23-year old Flying Officer John Roberts DFC of Brighton, South Australia who had taken the aircraft off from Waddington at 1904 hours. Born in Bowden, South Africa on 23 April 1920, he had enlisted in the RAAF on 11 October 1941. 'J-Jig' had been brought down in a collision with 'M-Mother' on 44 Squadron skippered by Flight Lieutenant Terry Fynn, which was shot down at 28,000 ft by Unteroffizier 'Hans' Müller of 1./JG301 and went into a steep dive before impacting with 'J-Jig'. Both aircraft still had their bomb loads on board and the two Lancasters broke apart in the explosion killing seven of the crewmembers on 'M-Mother' and everyone on 'J-Jig'. The impact was so great that Flight Sergeant James Macadam Benzie RAAF the 20-year-old bomb aimer on Roberts' Lancaster was later found dead in the wreckage of 'M-Mother'. He left a widow, Ivy Benzie of Etwall, Derbyshire. Only two complete bodies of Roberts' crew and five of Fynn's crew were found at Kornwestheim. Six other crew members were later buried in two graves to represent all remaining members of the two crews. Fynn, who had somehow survived, was captured and he spent the rest of the war behind the wire at Stalag Luft I, Barth. In Harare in a very humble flat just diagonally opposite the Catholic Cathedral, the mother of John Berrington heard on her son's 21st birthday, 18 December 1944 that he had died in the crash. John Fynn, Terry Fynn's brother was in training in Rhodesia to be a pilot when he got news of Berrington's death. It made him lose his nerve and he gave up training.[8]

Twenty Lancasters on 630 Squadron at East Kirkby had begun the night on the Battle Order for Stuttgart. Pilot Officer Leonard Alfred 'Barney' Barnes who skippered 'P-Peter' had recently celebrated his 24th birthday but tonight he had got off to a bad start. As usual he had put his lucky little teddy bear, which his fiancée, Merville, had given him, on the compass. But as soon as he started the engines the little bear fell to the floor because of the vibration. 'Barney' felt that this was a bad omen.

---

8. See *The Crew of Lancaster ND576 KM-M 44 Sqn, 15th March 1944* written and researched by Dorothy V. Ramser (May 2020).

Up until now his lucky mascot had brought nothing but good fortune. The first operations for this crew were to Berlin. During their first on 30/31 January whilst returning to England at 21,500 ft above 10/10ths cloud a Me 210 made a couple of passes at them over Magdeburg. Sergeant Thomas Austin Fox the 21-year-old rear gunner, better known as 'Freddie', a nickname he acquired after the famous jockey of the 20s and 30s, reported tracer incoming from the starboard quarter up. When he ordered a corkscrew to starboard he immediately sighted the 210 at 600 yards and he delivered a 2-second burst but no hits were seen and he made no claim. A minute later Fox sighted another Me 210 at 500 yards range. As Barnes began his corkscrew Fox and Sergeant James Henry Overholt the 20-year-old Canadian mid-upper gunner from Eastwood, Ontario opened fire with a 5-second burst, both noting strikes on the nose and engine of the enemy fighter. It broke away trailing smoke and was not seen again. The following morning the crew went out to assess the damage, only to see an unexploded cannon shell being removed from one of the fuel tanks. "Luck", said Barnes "was with us!"

At Stuttgart bombing took place between 2315 and 2329 hours from 20,000 to 23,000 ft. However, some on 630 Squadron at East Kirkby believed that the Pathfinders were late but that their markers had been quite well placed based on their bombing photos which showed the River Neckar and railway lines. Some crews encountered moderate to light defences in the target area whilst others suffered heavier flak and some reported intense fighter activity from Strasbourg and on the line of withdrawal. The crew on 'S-Sugar' skippered by Flight Lieutenant David 'Robbie' Roberts, a 21-year-old Londoner had the unnerving experience of seeing a bomber hit the sea on the outward journey and considered the bombing may have been scattered. Pilot Officer Cliff Rogers the skipper of 'K-King' noted large fires burning near the railway station and also east of the River Neckar. The crew of 'M-Mother' skippered by Pilot Officer Kenneth Watson Orchiston RNZAF of Beaumont, Otago and Pilot Officer Alan William Wilson RAAF of Murrumbeena, Victoria the 22-year-old captain of 'Q-Queenie', reported scattered fires. Orchiston would lose his life on 18 March 1944 aged 23, leaving a widow, Rose Elizabeth, of Ashburton, Canterbury, New Zealand. Wilson would die in action on the night of 6/7 June in the attack on Caen.

At 0130 hours west of Laon, 28-year-old Hauptmann Ludwig "Luk" Meister of 1./NJG4 who was flying a Bf 110G-4 from Florennes airfield in Belgium shot down 'V-Victor' on 630 Squadron. This Lancaster, skippered by 22-year-old Pilot Officer Kenneth Rodbourn, had been the first to take off from East Kirkby and was the second of two bombers Meister shot down that night in the Picardie region of France. (Earlier, at 0118 hours, 'B-Baker' on 100 Squadron at Grimsby (Waltham) piloted by 27-year-old Flight Sergeant Arthur Dawson, which Meister misidentified as a Halifax, had crashed at Bonneuil-les-Eaux, south of Amiens killing all seven crew. Dawson left a widow, Joan, of March, Cambridgeshire). 'Victor' went down out of control but eye witness reports suggest that the Lancaster was crash landed successfully at Besmé near Bourguignon-sous-Courcy in the Aisne, 23 km north-west of Soissons but then exploded due to a bomb which had failed to release detonating as the crew made their attempts to exit the aircraft. Kenneth Rodbourn left a widow, Monica Sheila Rodbourn of Shirley, Croydon, Surrey. Flight Sergeant Ernest John Philipson RAAF, the 25-year-old rear gunner, also left a widow, Mavis Elsie Philipson of Coburg, Victoria, Australia. Sergeant Richard John Easter, the 21-year-old flight engineer; Sergeant Albert Henry Wilkinson the 20-year-old navigator; Warrant Officer Alexander McCowan Freeman the 24-year-old Canadian bomb aimer; Flight Sergeant Frank James Hobbs, the 35-year old WOp/AG and Flight Sergeant Leslie Hall, the 23-year-old mid-upper gunner were the five other crewmembers who died.

Everyone on board 'P-Peter' on 630 Squadron were relieved to be making their homeward journey after Sergeant Malcolm Elliot (aka 'Ginger') Gregg the 21-year-old bomb aimer had dropped the bomb load on Stuttgart. 'Barney' Barnes had pulled sharply away from the city and all seemed to be going well but 24-year-old Oberleutnant Dieter Schmidt of 8.NJG1, flying from Ste-Dizier airfield in a Bf 110G-4 who had claimed a Halifax at Bonneuil-les-Eaux, south of Amiens at 0118 hours was determined to get a second kill this night. At about 0056 hours, 1 km southeast of Longwiller, France he carried out a typical Schräge Musik attack on the unsuspecting crew of 'P-Peter'. Inside the homeward-bound Lancaster there was a state of absolute carnage. The hydraulics had been destroyed and they were now a 'sitting duck'. Schmidt attacked for a second time from behind. In the rear turret 'Freddie' Fox could have known little of what happened next and he died in a hail of shrapnel. A keen amateur boxer, he had trained as a butcher under his father who was a master butcher. His son had wanted to be a pilot but at only 5 feet 1½ inches in height, 'Freddie' was considered too short. When he had joined the RAF in May 1941 he became a motor cyclist and despatch rider before training as a gunner.

Jimmy Overholt was by now lying on the rest bed, his oxygen supply to the nid-upper turret having been cut off. As 'P-Peter' was attacked again from underneath Overholt was killed. Barnes, unaware that his bomber was on fire put the Lancaster into a 360 mph dive from 12,000 ft to try to shake the fighter off. Finally the starboard engines were hit and set ablaze and the rudder controls were badly damaged and the order to bail out was given. Sergeant Ken Walker the 20-year-old flight engineer, Sergeant George Plowman the 21-year-old wireless operator and former leather worker from London, and Flying Officer Malcolm Geisler the 26-year-old navigator and a former Manchester salesman, jumped one after the other. 'Ginger' Gregg who had sustained shrapnel wounds to his hand managed to bail out with the rest of the crew. Upon landing (in a tree), he made his way across country and came upon a convent. When he asked for aid, he was given some food, but was subsequently captured and spent the rest of the war in Stalag Luft I. Two bodies were discovered in the remains of the aircraft which lay scattered in fields at Mont-sur-Courville close to Ste. Gilles (Marne) on the west bank of the River Ardre, 24 km west of Reims.

'Barney' Barnes, having realised that both starboard engines had been badly hit and were ablaze and the rudder controls badly damaged had repeated the order to his crew to bail out. Ensuring that his crew had left the aircraft, he prepared to make his own escape. However the Lancaster lurched to one side and he missed the escape hatch. After hitting the bombsight and knocking himself out in the process he regained his senses. The next thing 'Barney' remembered was falling through the air with an unopened parachute, his scarf slapping him gently on his cheek. Tumbling through the night sky he pulled his ripcord and floated to the ground, landing in a farmer's field obscured by a small wood. He found himself close to Dravegny in the department of Aisne. 'Barney' buried his parachute, Mae West and the tops of his flying boots and set off in a south-westerly direction. After about a mile he found another parachute hanging in a tree. 'Barney' dragged it clear and buried it but forgot to look for the name on the 'chute. (He discovered much later that it belonged to Ken Walker, who despite being wounded in the leg by shrapnel successfully evaded capture by walking for a week and covering approximately 70 miles until he came to La Ferte Gauche where he sought refuge in a farmhouse just outside the village. The farmer allowed him to hide and supplied him with food and aid).

'Barney' lit a cigarette and began his journey to freedom. In London on the morning of 17 March meanwhile, Merville his fiancée had pinned on her lapel the RAF wings brooch that 'Barney' had given her. As she bent forward the brooch fell to the ground. The clasp was still fastened! Merville wondered if this was a bad omen. But 'Barney's luck was in. Aided by the family of French patriot Leon Coigne and hidden for weeks despite the enormous risks which they faced, 'Barney' eventually was moved to Paris before becoming one of the very last to escape over the Pyrénées with the Comète escape line on 4 June 1944.

Ken Walker had managed to evade capture by travelling overland to a French town/village called La Ferte Gauche. Here he sought refuge in a farmhouse just outside the village where the farmer allowed him to hide and supplied him with food and aid for his wounded leg hit by shrapnel. He had travelled approximately 70 miles and walked for a week. Finally, in September, the Americans liberated the area where Ken Walker was in hiding. Both he and his skipper were flown back to England from Gibraltar. Upon his return to England 'Barney' completed the remaining 26 ops on the squadron. In October 1945 he and Merville were married and having disbanded 630 Squadron as adjutant, 'Barney' returned to his job as a printer for Glyn Mills & Co.[9]

Sixteen of the 17 Lancasters on 463 Squadron that had set out for Stuttgart returned over England and began letting down for landing back at Waddington. Pilot Officer William Alexander Graham of Queenscliffe, New South Wales had taken 'E-Easy' off at 1927 hours and successfully completed the operation. Whilst circling the aerodrome preparatory to landing, the 25-year-old pilot was involved in a mid-air collision near

---

9. See also, *Free To Fight Again: RAF Escapes and Evasions 1940–1945* by Alan W. Cooper (Airlife, 1988).

Branston, 4 miles south east of Lincoln with 'L-Leather' on 625 Squadron skippered by 26-year-old Flight Sergeant John Percy Bulger RCAF. All fourteen members on the two Lancasters were killed. Sergeant Henry Alfred Baxter, the 22-year-old wireless operator/gunner on Graham's crew left a widow, Vera Lilian Baxter, of Wandsworth, London. Flight Sergeant Alfred Samuel Humphreys of Gunnedah, New South Wales was 20-years-old. Bulger left a widow, Mary Margaret Bulger of Toronto. Jean Aiston, who was married to Sergeant Alfred William Aiston, the 35-year-old flight engineer, was living in Whitburn, County Durham in South Africa. One of Bulger's air gunners, Sergeant John Douglas Dadswell of South Shields was 19-years-old. Warrant Officer Class II Gordon Ivey, the 23-year-old navigator was from Ellscott, Alberta. The air bomber, 24-year-old Flying Officer Robert Roy Jones came from Medicine Hat, Alberta.

At Elsham Wolds, seven nerve-shattered airmen on 19-year-old Pilot Officer Bradfield Lydon's crew on 103 Squadron climbed out of their Lancaster after their harrowing trip to Stuttgart and made their way to de-briefing. Meeting fellow pilots, an exhausted Lydon said, "What a trip. I've never known such fighter activity. They must have had hundreds up against us!

How did you fellows cope?"

The replies were almost unanimous – "the quietest trip of their tour".

Cecil, a replacement gunner Lydon had taken with them for his first op was the only dissenter. They had been flying little more than half-an-hour when Cecil's voice on the intercom became incoherent and finally non-existent. Brian Lydon called him several times but there was no reply. Finally he asked him to flash his call light if he was receiving messages. This was duly done so the pilot instigated a battle code: two flashes meant a fighter was attacking, one that it had been shaken off. Almost immediately came two flashes and the navigator found himself suspended between the navigator's seat and the fuselage roof as Lydon took violent evasive action. Three minutes later it happened again and it was to keep happening all the way to the target and most of the way back to base. Cecil, it seems, had mistaken every Lancaster he saw for a Me 110!

All twenty-two crews on 166 Squadron that attacked the target, returned safely to Kirmington, although 'W-Willie' piloted by 20-year-old Flight Sergeant Victor Leonard Perry from Depford, London, was attacked by a night-fighter no less than seven times. 'He was on his first operational sortie too' wrote 'Jonah' Jones. "The rear gunner replied to the first three attacks, then his guns jammed and he was killed in the next attack. The mid-upper gunner also was wounded. The aircraft was extensively damaged but the pilot brought it back and crash-landed at base."[10] Perry and five of his crew were killed a few nights' later, on 26/27 March on Essen. Judging by the crew reports at interrogation after the Stuttgart raid, Jones had doubts whether the attack had fallen within ten miles of the target. Adverse winds had delayed the opening of the attack and they may have been the cause of the Pathfinder marking falling back well short of the target.[11] Some of the early bombing had in fact fallen in the centre of Stuttgart at Vaihingen and Möhringen but later German reports showed that the bombing did indeed stray out through the south-western suburbs into open countryside and a number of neighbouring towns were also hit. A total of 86 people were killed and 203 were injured.

In all, 53 air raids were launched against Stuttgart by RAF Bomber Command and the US Eighth Air Force who dropped 20,000 high explosive bombs and 1.3 million incendiaries on the city by the time hostilities ended in 1945. As a result 57.7% of all buildings in the city were destroyed, 4,477 inhabitants were killed, 85 citizens disappeared and 8,908 more people were injured. The heaviest raid took place on 12 September 1944, when RAF Bomber Command dropped over 184,000 bombs – including 75 blockbusters that levelled the city centre and killing 957 people in the resulting firestorm. Günther Schaile, a twelve-year-old boy who lived with his mother in Steinhaldenfeld and who had lost his father in Stalingrad in 1942, was always surprised that new holes were always being dug next to the cemetery, apparently graves for many people. He soon learned that they would fill up quickly. Young people and older men of the Volkssturm had to recover people buried in the ruins of houses. 'The entrances had to be shovelled free with a spade and stones pushed away with your hands but most of the time there were only dead people in the cellars. The phosphorus bombs, the sea of flames – people had no chance. The bombers destroyed the heart of Stuttgart with

---

10. *Bomber Intelligence* by W. E. Jones (Midland Counties Publications, 1983).
11. *The Bomber Command War Diaries: An Operational reference book 1939–1945* by Martin Middlebrook and Chris Everitt (Midland 1985).

tens of thousands of explosive and incendiary bombs. The new castle, the town hall, the area from the castle square to the Wilhelmsbau, from the Leonhard church to the Katharinen hospital were only smoking rubble. In the four nights of July 1944, Ostheim, Gablenberg, Botnang and Feuerbach were also attacked, the hail of bombs reached Untertürkheim. A total of 884 people died and 100,000 became homeless. Among the dead were sixteen high school students who helped man an anti-aircraft position in Degerloch."

Chapter 7

"For Those Who Wait"

'Ray saw the missing five appearing one by one. He saw Joe returning on a ship and rushed to meet him at the docks. He saw him as the ship entered the harbour and waved. Joe waved back and smiled. They looked at each other as if they were strangers, but only for a moment. Joe walked down the gangplank and Ray took both of Joe's hands and shook them gently.

'For a second they hugged each other. The spectators laughed, but not 'all laughed, for many understood. Together, they walked away and Ray whispered: 'It's great to see you, Joe.'

'You too Ray.'

'What of the others?'

'They're on their way back.'

'I'm so glad, so glad' and the dream came to an end.'

For *Those Who Wait: A Story for Bomber Crew Next-of-Kin* by Flight Lieutenant Surender Lal Berry DFC RAFVR. (Morris & CO, Dublin, 1944). 'Sammy' Berry, a Lancaster navigator on 622 Squadron at Mildenhall, was born in Calcutta of Indian parents but lived nearly all his life in England. He was on the thin side and had one of those faces which women look at and sigh. His hair was jet black and he had green-blue eyes. His father was prominent in Indian business circles, while his mother was closely identified with the movement for the emancipation of Indian women. Most of Berry's air training took place in Canada.

In his book he calls himself "Raymond Clive" – 'a 20-year-old whose manner gave the impression that he was older' and calls his pilot, Flight Lieutenant James Andrew Watson RCAF, 'Joe', describing him as having 'a face so inconspicuous, so very ordinary and average, that he could be mistaken very easily for someone else. He was quite handsome except his lips were a little too thick. His hair was mousey and this, with his grey-blue eyes and slightly stubby nose, gave him a very fresh and clean-cut appearance – the freshness of youth.' Born in Hamilton, Ontario,

9 June 1922, the son of Robert Scott Watson MC and Mary Kathleen Watson, he was the old man of the crew at 21 years of age. A former student of Westdale Collegiate, he was scoutmaster at Melrose United Church and a member of the staff of Camp Onondaga. Watson won his wings and commission at No. 5 A.N.S Brantford on 23 October 1942 and was posted overseas before the end of the year. He became a flying officer in April 1943 and within two months was promoted to the rank of flight-lieutenant.

Spring 1944

As the transport brought them from the railway station to the camp, they saw the monsters which they were going to fly. They looked such giants, these four-engined aircraft. So impressive that it was inconceivable to them that they alone were to fly them; they seemed so large. They thought the designer must have possessed inspiration as well as genius, as they were filled with an impression of exquisite beauty and wonderment which left a permanent mark on their memory. They felt these 'planes were part of the future, the great things which were yet to come. Their eyes filled with admiration and remained as if glued. They found it difficult to look away from these fascinating monsters, the Lancasters of Bomber Command.

Training was in full swing and they flew and flew. They studied this magnificent aircraft to its most minute detail, until they were thoroughly familiar with every particle. The more they learned, the more they liked thus wonder of wonders and felt delighted that they had been chosen to fly a Lancaster. But they also learned that mastering the four-engined bomber was not easy. There were more tasks and difficulties. There was so much to learn and so little time in which to absorb it. It was work, work and work.

The crew looked at the Lancasters, which had ceased to be a novelty. They knew the aircraft so well. But they could not help admiring them that afternoon as their shining perspex reflected the sun and added to their lustrous beauty. The bomber had ceased to be a 'new toy' to them and they had much to do. 'Come on' said Joe, shepherding them, 'we've all the interviews yet. Let's get them over.'

Then came the 'Battle Order' and for the first time they saw their names on it as a crew. They tested the aircraft and some of the ground crew flew with them. The ground crew loved to fly, if only as passengers. They had told Joe that they were ground crew because they had been rejected for air crew and felt it was the next best thing. In one way or another they had not come up to the requirements of air crew.

Then came the briefing and they were given all the details of the mission and the strategy of the battle. They had their first conception of the ingenious planning: the brains behind the operation, the magnificent and intricate organisation, as detail upon detail was covered and then at last, after arduous planning, they were ready.

The only indications of the nervous tension present were the amount of tobacco smoke and that some flyers were eating or chewing.

The crews entered the waiting transport which was to take them to their aircraft. When the bus was full it raced around the aerodrome and stopped at each aircraft. The crews spoke little but were always ready to laugh at anyone who jested.

The transport stopped for Joe's crew and they were by their aircraft. They made their last-minute preparations and everything was in order. The time to take off was an hour hence and they laughed and smoked cigarettes whilst waiting to take off. The CO came to see them and spent a few minutes chatting with each one. Then the M.O. had a few words with them and then the Leaders and so on. Everybody was talking to someone.

At last the moment came. They grinned and signalled 'Thumbs up' to their large audience which had collected. A cheery word to their ground crew, who signalled 'Thumbs up' and they were ready in their positions. The engines roared to life, the lights flashed on and in a few moments they had taxied from the dispersal point and were ready for the take off. The engines roared into their full power and their speed increased and increased, until at last they were off. They circled and climbed with the other aircraft and felt reality was truly with them for the first time. They knew they were on their way to the Continent, heading straight for battle, for danger and perhaps death.

Things were quiet until they reached the enemy coast, which Ray announced. The enemy greeted them with flak, which had a fascinating and intriguing beauty. A beauty full of evil, for it wanted to lure them to their doom. They battled on and all was well until they reached the target and were dropping the last of their load. As soon as it had gone, Johnny said, quickly but calmly: 'Aw-aw-weave to port quick, 'Skip'!'

Before they realised it they were slipping to port and then they saw the reason. The very spot where they had been but a few seconds before showed great illumination, as shell after shell exploded.

Then, on their way back Tommy suddenly came forth with: 'Hallo Skip, Jerry on starboard beam, two thousand yards.'

'Carry on Tommy' replied Joe.

'He's on a parallel course – I don't think he's seen us.'

'Keep your eyes open Mickey' said Joe.

'OK Skipper.'

'He's turning now towards us and climbing' commented Tommy. 'He's still turning – on a reciprocal course now and going away from us. Over to you Mick.'

'OK Tom I see him. He's still flying away from us. It's a Heinkel. Still further away – he's just a dot now – he's gone.'

'OK Mick but it might be a trap.'

'Yes Skip.'

The gunners remained silent for they saw no more on which to comment.

Then at last Ray said: 'You should see base Skipper, slightly on the port bow now.'

'Yes I see it Ray.'

'Well we're home boys,' rasped Mickey.

'Oh boy we're operational!' laughed Dicky.

'Let's get down before we go into hysterics' said Joe coldly.

The crew then restrained their emotions until they had landed and parked the aircraft. Then pandemonium broke loose as they rushed out of the aircraft, but Ray stayed behind to tidy his desk and place his charts, logs and instruments in his bag. He followed them and Mickey took the bag as he stepped down the ladder.

'Here, give it to me Ray.'

'Thanks, Mick.'

Well clear of the aircraft, they lit cigarettes.

'I wonder why that German didn't go for us' laughed Dicky.

'Either he didn't see us or was scared.' Wish we could have gone after him, Skip,' rasped Mickey.

'I know, but it's not our job to attack aircraft. If we went chasing, we'd never get back.'

'Your chance will come, Mickey,' grinned Johnny.

'Johnny, you look as if you've just got up.'

'You look as if you've been in bed with a woman for a week,' he retorted, with a laugh.

They all laughed as they noticed each other, for all looked the same – worn, tired and as if they had a happy hangover.

'Gee it's cold,' grinned Johnny.

'What are you grinning at?' laughed Joe. 'I know it's cold, but what's funny about that?'

'Then why did you laugh?'

They all laughed heartily, as they realised they could not help it.

'I guess we're just glad to be operational' smiled Ray.

'I guess we're just happy to be back' laughed Titch.

The transport arrived and they hurriedly scrambled in. Other crews were already in it.

'Come on' they jibed, 'let's go, eh?'

This transport was vastly different from that which had taken them to their aircraft, although it was the same van. Everyone was chatting and laughing nonsense, with a considerable amount of noise.

They entered the briefing room, where they were to be interrogated. At the doorway stood the padre with a tea urn at his side. He poured out tea and handed it to them as they came.

'Cigarettes here' he smiled. 'What about a tot of navy rum?'

'Not for me thanks padre' smiled Ray.

'I'll have some in my tea please' said Mickey.

The intelligence staff quietly sat at their tables, which were each surrounded by seven vacant chairs. They patiently waited for the crews to seat themselves.

It was noisy. Hysterical laughter and chatter filled the room as Ray carried his log and chart and the crew sat down at a table.

Then they reported the trip in detail, all speaking in turn. They realised that possibly what they may well think as useless information sometimes was the very link which completed a jig-saw puzzle to the intelligence staff.

After interrogation the crew deposited their flying kit in their lockers. These were in the adjoining room. Then they made off to the night mess for breakfast.

'Gosh I'm hungry' said Dicky.

'I guess we all are' commented Joe.

The crew shared a table as they ate their egg and bacon.

Tommy said: 'That's the one thing I appreciate about it all. We eat together as a crew.'

'Don't forget the bacon and egg' said Johnny.

'You're a wolf with food as well as women' laughed Mickey.

'Skipper, do you hear what he's saying about me?'

They all laughed as they ate and, after a cigarette, Joe and Ray remained seated, while the others made off to bed.

'Aren't you two coming?' queried Johnny.

'In a minute' replied Joe.

'Well I'll go on – goodnight.'

'Goodnight.'

'Cigarette, Joe?' offered Ray as they sat. 'Have a Canadian one for a change.'

'Thanks. Where did you get these from?'

'Got a parcel yesterday.'

'Good show; that means the crew smokes again.'

'Well' smiled Joe, 'what did you think of it?'

'We've a lot to learn. We should go over it in detail.'

'Yes, we'll have a crew discussion tomorrow. Let's go to bed.'

'Are you tired?' asked Joe, as they rose and walked to their mess.

'A little.'

As they undressed, Ray said: 'You know Joe, if that's an easy trip, as it's supposed to be, I wonder what a tough operation is like?'

'It just shows, you never can tell – there's so much luck. You might catch a packet anywhere. There are no rules to the game. No one can say you won't be worried here; there's nothing there.'

'Yes I know, we've got a lot to learn.'

'We have and we will go on learning until our last operation.'

'Goodnight Joe.'

'Goodnight.'

In the morning Joe awoke rather early and was surprised to find Ray awake.

'What's the matter – couldn't you sleep?'

'No.' Ray smiled. 'I suddenly woke up and I wondered if I had been dreaming. Last night was so unreal.'

'I know – just like a nightmare. We'd better get up.'

'Yes, it's time we did' and they scrambled out of bed and raced to the wash basin and Ray laughed, for it was the first time he had beaten Joe to it. He was usually content to doze in bed until the last possible moment.

'A few more ops and you'll be perfect,' cracked Joe and Ray laughed as he washed. He saw Joe looking at him through the mirror.

'Seeing a ghost' smiled Ray, with soap all over him.

'I was just noticing your back, with all its ripples; you're quite wiry.'

'Go on, I'm skinny as a pole.'

'And I'm fat as a barrel' laughed Joe, patting his stomach, which showed tendencies of protruding in azimuth.

'It's the beer' laughed Ray, rinsing himself. 'I must be getting older.'

'You are older.'

'So are you.'

'We are both a couple of old fogeys.'

Joe imitated age by hunching his back and uttering a lamb-like bleat; they laughed.

Their day began with a signal that Bomber Command was on operations that night. Then follows the 'Battle Order, or 'Crew Lists,' informing the air crew who will be in battle. They do not know where they are going – no one knew.

The work began: they make their test flight. Everything must be in perfect order.

When they return, the ground crews work at a furious pace. By evening the aircraft must be faultless.

They have a hot meal and the work continues. They are briefed and the target is made known – it is Mannheim. They plunge into arduous detailed planning. Those are few words to describe a briefing, but Ray's conscience did not permit him to write of the briefing in detail, for he cannot do it without revealing all kinds of secrets, but it was really inspiring to be briefed.

The intelligence staff informed them what factories they are bombing and why. They told them what part it will play in denying the German forces' supplies and thus weaken them. They told them why the route was the best one. Of the tactics, different heights, speeds, descents, climbs at various points, to add to their safety. They informed them most accurately of the German's defences – in fact, a hundred and one details which never ceased to amaze Ray. While he listened to it all, he was filled with admiration for the people who risked their lives so gallantly to provide this information and make the intelligence service the most superior in the world. They were the real heroes, for there is no glory in being captured as a spy.

Then Jock the 'pukka' meteorologist told them what kind of weather to expect. He was so convincing and so invariably right Ray often felt he has a private intelligence service.

Whilst Ray was working away on his pre-flight plan and listening at the same time, Dicky and Harry too, made their log preparations. The remainder *paid more attention and listened more intently, for the rest of them were the eyes of the aircraft.*

Harry was informed of his signals data, Dicky of his petrol, the amount in each tank; the safety weight of landing, if by mischance they had to make an early return. Johnny was informed of his bomb load – how they are displaced over the aircraft and how he must release them that night. While for an hour and a half all this was going on and Ray was sweating away working out a navigation plan to the smallest detail, Joe was aware that outside their briefing room everyone was working as never before.

Finally, the CO summed up the briefing and told of the major plan. They were to bomb zero plus two. Zero hour is 2300 hours, when the attack opened. He told them that a dummy attack was to be made on Frankfurt at zero minus 10, some bombs would actually be dropped and crews must be sure of their position at that time. This was to lure the majority of the German fighters away while they were over the target. Then to fool the German's warning equipment of their approach, hundreds of training aircraft were flying to a point, but a, few miles off the enemy coast on a Northern route. They were given a zero hour to be at this point before turning round and coming home. They would arrive at this point ten minutes before they penetrated the German defences on the Southern route and all the fighters would be by that time concentrating on the North, whereas attackers would be sneaking in on the South. They will discover them eventually, but it would be too late for them to do anything on a grand scale. For every ten fighters they might have had they shall only get one – all this assuming that the plan worked well.

The CO concludes by telling crews that Mannheim was extremely well defended. The German had a large belt of searchlights around the city, making a circle of fifteen miles in radius. He had thousands upon thousands of guns there and masses of radar equipment to plot them if they did happen to be over cloud. They knew it was a tough target and Joe memorised the route so that when Ray gave directions he understood what was happening without a moment's hesitation.

Biologists have been known to admire the amazing instinct of the ants, as they are found in their thousands to be working in perfect harmony with a definite aim: but they did not know of the genius, the magnificence, of the organisation of Bomber Command.

The pace of activity increased to an overwhelming tempo as everyone prepared for the gigantic battle. The armourers were seen with their bomb trolleys, as they loaded the aircraft. The petroleum wagons were refuelling. Each aircraft was full of electricians, technicians and engineers. Everything was checked and rechecked. Everyone felt that nothing must be allowed to add unnecessarily to the danger for the air crews.

Finally, all was ready and they collected their rations. Crews were taken to their aircraft. The engines were quickly run up, checked and the last-minute preparations completed.

They left the aircraft and, at a safe distance, quietly smoked and chatted as they waited.

There was nothing more they could do now, but wait…

There was only the sound of a cool breeze which seemed to bring a message. On and on it whispered and there was a feeling that the calm was only a guise. Something frightening, fantastic, was about to happen. The sun bade farewell and the birds were silent. The shadows of night began to creep on. The golden fields were lustrous in the full glory of the twilight and stretched on and on, as far as the eye could see. Here and there a field was disfigured by the scars of concrete runways on which the aircraft rested, looking like fabulous beetles. The plains of East Anglia might have been uninhabited, for there were no people moving over the fields.

The stars appeared one by one, breaking the spell of twilight in order of their magnitude. The moon had not risen and the heavens were glorious.

The gentle breeze ceased its whisper and all became deadly quiet. Then the change came – in a flash. East Anglia was no longer the quiet pleasant countryside. It sprang into vivid, raging life. Lights flashed like gigantic neon signs and there was a terrifying roar as the monstrous beetles came to life.

A few minutes and the sky filled with hundreds upon hundreds of four-engined monsters and the noise reached a maddening crescendo, which made thought unbearable. The air trembled and everyone was filled with hysterical vibrations. Still the uproar continued, immersing everything to the depths of despondency. Then gradually the noise became less, changing into a gentle hum as the 'planes all went towards the South. The countryside resumed its pleasant atmosphere as the gentle hum and the lights disappeared into nothingness.

As the planes set course, they formed a stream of red, green and yellow lights, which floated merrily along, like moving stars, twinkling a challenge to the planets above.

It was a beautiful sight for those on the ground, but Joe could enjoy it only for a second. The thousands of lights constituted a danger to each other – the danger of collision. No-one could relax; constant vigilance was essential. Soon these lights would go out and night vision become of supreme importance.

They were routed over London.

Ray said: 'Position 'A', just past London, five minutes, Skip.'

'OK Ray' Joe replied.

As they flew over London, Joe felt it was an inspiring sight. It looked like any other town – blocks and blocks of buildings – but the Thames, with its many bridges, added a silvery twinkle and the dome of St. Paul's was majestic in its might. London looked alive, as he saw millions of spots of light and I realised they were the torches of the pedestrians thronging the streets. The lights flashed at them.

Mickey said: 'Look, Skipper, they're flashing 'V'. Boy, what a sight!'

'I see them'. Mickey and Joe returned the compliment by flashing on the navigation lights.

Ray felt this was the feeling – the message from the people of a very great city.

Gradually the navigation lights were switched off. All was dark, for the twilight had ended. They were swallowed into the night and the heavens were so serene and peaceful that for a moment Joe had the illusion that the world was at peace. But this was no time for illusions. From now on there was not a moment, not a second, to relax as they raced into the greatest battle the world has ever known, the fierceness of which was inconceivable to the most imaginative mind. They could not relax in the midst of a thousand aircraft, each one veiled by a curtain of darkness and only to be seen at close range, when collision was so easily possible if they were not absolutely on the alert. Good night vision was very important to safety. To be drowsy was to commit suicide.

London was out of sight behind and Ray said: Skipper, you're at position 'A'; turn now on one two zero for position 'B' on our coast.'

'OK' and Joe turned immediately.

He saw the coastline of England loom up ahead, in a few minutes they would be over the sea and he had the feeling of leaving safety and piercing the unknown. They spoke little, for Joe did not encourage idle conversation, but it was really unnecessary, for they were fully aware of each other: in the atmosphere there was a spirit which needed no expression. It made them each feel part of one being, instead of seven individuals.

Ray said, 'you are at position 'B', 'Skip', turn zero eight zero for position 'V' on the enemy coast.'

Joe turned and as he levelled out Dicky said: 'Your boost and revs, Skip.'

'OK Dicky' and Joe adjusted them to his satisfaction. *Dicky always seemed so concerned about the engines and fuel consumption. I don't really know what I'd do without him.*

Then Ray said: 'Enemy coast in fifteen minutes Skip.'

'OK Ray.'

Joe knew in a few minutes he would see the enemy coast ahead and asked the crew how they were. He made a practice of doing this from time to time, after a period of silence.

'How's everything Mickey?' Joe asked.

'All right Skip.'

'OK Tommy?'

'Right as rain.'

'All right Harry?'

'High as a kipper Skip.'

'Johnny?'

'Get up them stairs.'

It always heartened Ray to hear their wisecracks.

He saw the enemy coast appear through the veil of darkness and the feeling in him of piercing the unknown became stronger and stronger. He knew it was fear, mingled with uncertainty. It reached a climax as they crossed the coast and then suddenly all the feeling vanished, for the real battle had begun; the battle of wits, alertness and brains – not forgetting the inevitable fact of luck. Ray felt here was the grim reality of war. Nothing was said of their feelings – that was a personal affair and each understood the other's without any indications of their hopes or fears.

Ray may well have thought the scene was a firework display as the sky filled with rockets and shells. But they were so deadly it did not leave room for any appreciation of beauty. He felt a terrifying suspense when they saw a gun flash on the ground and knowing that, in a few seconds, the shell will burst, they hoped far away, but often it was too close. It was an awful moment when a red-hot rocket seemed to be heading right for their plane. Joe banked, Joe dived. He twisted and still it followed so persistently; then, to his relief, it exploded before reaching them. Joe weaved to avoid the searchlights combing the skies, for to be caught and coned was asking for trouble.

Throughout all this Joe was constantly aware of the thousand aircraft around and he ensured that every possible pair of eyes continuously searched the skies, ever peering into the darkness, vigilantly alert.

Joe felt a sudden jolt, which seemed like a burst of flak. He made a quick check up – everybody's all right – and Joe realised with relief that it was only the slipstream of an aircraft ahead.

Mickey, I think, is the coolest guy I have ever met. His courage overwhelms me as he will make us laugh at the most fearful moment. He reports any close burst of flak behind us.

He said: 'Skipper, flak two miles dead astern.'

'OK, Mick.'

'Skipper, flak one mile astern.'

'OK Mick' and Joe weaved to avoid the next burst, for he knew that 'Jerry' was plotting them.

'Skipper, flak half a mile on the port bow.'

Then Mickey said: 'Skipper, flak half a mile on the starboard bow.'

This gunner was clever and Joe tried to fool him by not weaving at all. He was fooled for a moment, Ray knew, when Mickey said: 'Flak five hundred yards on the port bow.'

The next few minutes were very trying as Joe weaved and then flew steady and then weaved. I could feel the sweat running down the sides of my face and sticking to my helmet.

'Flak four hundred yards dead astern.'

Then Mickey suddenly said: ' Skipper, the flak's right up my aunt fanny: let's get to hell out of here.'

They were through that patch of flak and Joe knew everyone was chuckling, though their microphones were switched off. They were very lucky; that final burst had jolted them and they had been hit by odd pieces of shrapnel, but fortunately no one was hurt and no serious damage had been done.

Poor Ray was continuously sweating away at his desk. This weaving did not make his task any easier and he continued to give Joe new directions to keep them on the route and also to ensure that they get there on time. The success of the tactics depended upon everyone being there on time and they must not be early or late. That in itself was as dangerous as anything, for they would be victims to the German's entire forces.

So they battled their way through and the hordes of enemy night fighters attacked. On and on they came. Ray felt it became a 'survival of the fittest'.

Mickey spotted one: 'Enemy aircraft one thousand yards, port quarter up.'

'Carry on Mickey' Joe acknowledged quickly.

'Closing into eight hundred yards – ready to corkscrew port – six hundred yards – go!'

Joe corkscrewed to port while he heard the rear turret firing its burst of fury, at the same time he threw the aircraft all over the sky, first down in a dive and then up in a steep climb, turning at the same time: the firing continued throughout.

For a second there was dead silence – it was a terrible moment, for Joe knew all wanted to ask: 'Mickey, are you all right?' Were it not for crew discipline they would all be speaking at once.

Joe swallowed hard and was about to speak with a well-practised voice, for he felt he must always try and appear cool, but Mickey said in a flash: 'It's OK Skipper, I got him.' Once more his eyes scanned the sky. They all said such things as, 'Jolly good show, Mickey!' and that was all, for there was no time for great rejoicing – the battle was still raging in all its fury.

It was inconceivable that the combat was but for a few seconds, which meant life or death to us or the enemy. Yet in those few seconds, while Joe was taking evasive action and Tommy and Mickey were tiring, Dicky was carefully watching the petrol gauges in ease of hits, Ray was noting position of attack, the time, direction and type of aircraft – in fact, all possible details: Harry remained at his set as if nothing was happening – it was necessary regardless of such incidents as these – and Johnny in the front turret manned his guns. The combat was over and until the next one came, they were back to their normal business of reaching the target on time – the normal business of battling through all the German's defences. On they went, hour after hour.

Time was relatively strange in an aircraft. Sometimes an hour went like the wind and a second appeared to drag for an hour, especially in a combat. So much happened in those vital seconds.

Mickey said: 'I saw him hit the deck in flames.'

'OK Mick,' Joe replied and knew that when other crews saw it they wondered, 'Ours or theirs?' for that thought always crossed the minds of each crew when an aircraft was seen to go down. At night it was very difficult to tell. Unless it was very close they would not know definitely until sometime later, when it had ceased to be significant to their minds, being away back in the past. Each day was like a year.

Ray knew zero hour was approaching.

'Mannheim in fifteen minutes Skip' I said.

'OK, Ray.'

Johnny then prepared to carry out the most important task of all; the object of the mission – to bomb and obtain direct hits on the target.

'Ready for the wind Johnny?'

'Yes Ray.'

'It's two five zero, forty.'

'OK thanks.'

'Skipper, we bomb zero plus two' said Ray. 'One minute to zero hour.'

'OK Ray.'

'Zero hour now; the first flare should be going down now.'

'It is, Ray.' Joe held Ray in high esteem, for there he was, shrouded by black curtains, able by his art, his science of navigation, to tell the crew what to expect. Ray knew it always amused him why it impressed the crew, for he felt that when one mastered an art such things did not appear uncanny.

Johnny was ready.

Then came a great moment of exceptional danger. The target had been revealed and the German was amazingly quick and clever. Hundreds upon hundreds of blinding searchlights filled the sky, already peppered by hundreds, maybe thousands, of shells and rockets. It seemed impossible to get through the searchlights and fury of fire; that terrifying barrage of red-hot metal. A steady run was necessary and it was a most vulnerable moment. In addition to the flak, swarms of night fighters began collecting above, ready to pounce and swoop down on us and other aircraft making then runs, taking full advantage of these most vital moments.

'Left, left – steady' said Johnny. The bombing run had begun.

'Right – steady!' he said sharply. His very tone indicated the degree of turn required.

'Bomb doors open' he directed.

'Bomb doors open' Joe confirmed.

A minute dragged like an hour at such a time – Ray could feel his heart pounding – were those bombs never going? Steady' and the seconds dragged – a watch meant NOTHING.

Then: 'Bombs going' and at last, 'Bombs gone'.

The bombs had gone, the camera clicked and, if all had gone well, they would have a picture which the experts would investigate and be able to tell them if they had obtained direct hits. It was a prize Joe treasured and of which he felt proud, for it meant that their mission was a complete success.

The target had been bombed, but the battle was not over. The climax remained as they battled their way over Mannheim's defences and were continuously being chased by enemy fighters.

The activity below appeared as the greatest pyrotechnic show in history. By comparison, the American Fourth of July displays paled into insignificance, but there was not time to appreciate it. Ray dared not look for more than a second. Mickey and Tommy dared not look at all, for fighters were liable to pounce that very second. One second late was too costly in this battle of the 'survival of the fittest'].

Up there in the stratosphere, where oxygen is necessary for existence, Ray felt it was easy to yield to temptation and take a good look, but the reality of bloody battle did not allow him time to brood on such things. They had yet to fight their way out of the heart of Germany and the German was doing his best to see that they were not going to be successful.

The battle continued and Joe was constantly aware of the thousand aircraft of which they were in the midst.

'Hallo Skip' said Mick, 'Lancaster port quarter down eight hundred yards.'

'OK Mick' Joe replied, as he turned away. That was much too close.

'OK Skip, you're clear now – resume course.'

'Right-ho Mickey.' Joe tried to be jovial. He felt cheerfulness of speech was essential for maintenance of spirit; it was as important as coolness of manner. No matter if his heart was pounding a hundred and twenty to the minute, cheerfulness and calmness formed the combination desired.

There was a certain exhilaration Joe felt, in the knowledge that they were on their way home. If time did not drag, they would soon be safe and sound.

Then it came without any warning:

'Fighter, fighter, starboard go!' came from Tommy and, before Joe realised it, he had automatically dived.

Tommy and Mickey were firing: their guns could be heard on the intercom. It was a terrible moment when tracer shot past the aircraft. This German was more vicious and persistent. On and on he came, with their tracer piercing into him and his into them. Who would it be? Then came the choking and sputtering from the starboard outer engine and it was useless – shot to shreds. Joe felt an awful fear that the petrol tanks would ignite and explode as the tracer poured in. Suddenly it all ceased and there was dead silence for that second which seemed like an hour.

'It's OK Skipper; he's gone' said Tommy. *'I hit him but I didn't see him go down – did you Mick?'*

'No, I lost sight of him.'

They claimed that as damaged and now their crippled Lancaster flew on with three engines. Joe made a quick check up with the crew – they were all right. Ray's lips moved in silent prayer and he thanked God.

They were lucky that time. Ray wondered at this factor of luck. He thought of it as a mystical quality which is intimately woven with the very mystery of life itself and must forever be merged in beliefs and faiths. Merged with fate, whether it is to be life or death, today or tomorrow.

Joe felt thankful for such a great crew and his thought turned to a lecture he had once received. He remembered it very well. The screen had said: 'Team spirit is essentially the pre-requisite foundation for maximum efficiency, which may mean life or death, success or failure. This team spirit inevitably evolves all into oneness of a crew, with the skipper at the head.' But he could not see how any one of them was singularly more important than the other.

Hour after hour the battle raged on. At last the enemy coast appeared; which gave Ray a feeling of exaltation.

'Aircraft shot down on port beam' said Tommy.

'OK' Joe replied and Ray noted all the details.

He thought: 'Ours or theirs? If it should be ours? Poor devil – to get so far and catch a packet here.

The enemy coast was behind them, but the battle was not over. It would not be until after a safe landing. The treacherous German was persistent and tried to sneak into the stream of aircraft, hoping to find them off their guard.

Nevertheless, Ray felt there was a tremendously joyful feeling as they saw the welcoming shores of old England.

Then to base and in a few minutes they were circling to land. Joe knew they were all very tired and tempted to relax, but they must not. He did his best to muster all efforts and concentrate on making a good safe landing. It worried him, for he found it difficult with a crippled plane. But the crew all helped him.

Mick said: 'Come on Skip, see how high you can bounce me.'

'I'll log the number of times you take off and land' said Ray.

'Let me do it' said Johnny.

'What about the poor wireless op?' said Harry.

'Remember I'm married Skip' said Tommy.

Dicky sang, 'Skipper's going to bounce us – Skipper's going to bounce us!'

Lancaster ED803 on 467 Sqn RAAF at Waddington which went MIA with Pilot Officer Jack Mitchell's crew on the Magdeburg operation on 21/22 January 1944. There were no survivors.

419 'Moose' Sqn RCAF taxi out at Middleton St. George. Lancaster KB711 went MIA with 23-year old P/O John Crawford McNary and crew on the night of 1/2 May 1944. McNary and his wireless operator P/O John Louis Chartrand were killed.

Pilot Officer Kevin Harold McKnight's crew on DV274 JO-R on 463 Squadron RAAF, that went MIA on 25/26 February 1944 on Augsburg. Flying Officer Stanley Alfred Isham, navigator, is 2nd from right. The crew were on only their 7th operation.

Lancaster LM486 on 300 'Masovian' Sqn, which went MIA with the loss of all seven men on S/L Kurowski's crew on 24/25 April 1944 on the operation on Karlsruhe. It is likely that they went down over the sea.

Lancaster DV389 'X-X-ray on 101 Sqn, which went MIA with Flight Sergeant Wilfred Albert Sullock and crew on Aachen on 25 May 1944.

Crew of Lancaster III ND758 'The Bad Penny' on 115 Squadron at Witchford in July 1944. L-R: Sgt Joe Hayes, rear gunner; Sgt Bernard Payne, WOp; P/O Eric Wilkin, mid-upper gunner; Flt Lt 'Mac' McKechnie, pilot; F/O Frank Leatherdale, navigator; Sgt Ken Denly, bomb aimer' Sgt Arthur France, flight engineer.

Their majesties King George VI and Queen Elizabeth with HRH the Princess Elizabeth escorted by the Stn CO, G/C Sims with aircrew waiting the call to attack a V-1 site in Northern France, RAF Witchford, 5 July 1944.

Lancaster takeoff.

F/O Jack Kaiser RCAF and crew on 90 Sqn at Tuddenham. From left: F/Sgt 'Bill' Wooldridge, flight engineer; 'Pete' Campbell, gunner; Ross Meggason RCAF, bomb aimer; Jack Kaiser; 'Jimmy' Ginn, wireless operator; Clare Telford RCAF, rear gunner and Howard Keon RCAF, navigator. (*via Mike Wooldridge*)

Ground crew carrying out engine repairs on a 419 'Moose' Sqn RCAF Lancaster X at Middleton St. George in 1944.

Armourers back a tractor and trolley loaded with a 4,000 lb 'Cookie' and incendiaries under the open bomb bay of Lancaster Mk.II LL725 'EQ-C' on 408 'Goose' Sqn RCAF at Linton-on-Ouse. LL725 was lost with all eight members of P/O John Harold Alexander McCaffrey's crew over Hamburg on 28/29 July 1944.

The Captain's Fancy (NE181) on 75 Sqn with Francis Woods' crew in August 1944 when this Lancaster had flown 39 ops before joining 514 Sqn in July 1944. Back row L to R: Arthur Taylor, wireless operator; R. Johnson, navigator; Francis Wood; Les Hurcombe, air bomber. Front row, L to R: Sergeants' Woolley, mid-upper gunner and Mahoney, rear gunner; Les Gibbs, flight engineer. (*Kevin King/Alf Gibbs*)

Damage to the Opel Works at Rüsselsheim caused by the night raid on 12/13th August 1944 by a force of 297 bombers which included 191 Lancasters, 13 of which were lost. Most of the bombs dropped fell in open countryside south of the target.

Lancaster B1 NN742 IQ-U on 150 Sqn seen here passing Lincoln Cathedral was part of a batch of 100 Lancs ordered from Austin Motors, Longbridge and powered by Merlin 24 engines. Delivered to 150 Sqn in October 1944, it was broken up for scrap in May 1974.

The Fleet Air Arm attacked the *Tirpitz* when she was in Altenfjorden during the spring and summer of 1944; dive bombers from aircraft carriers obtained direct hits and damaged her but in the summer of 1944 the *Tirpitz* was repaired and fit to go to sea.

'Tallboy' 12,030 lb 'Tallboy' deep-penetration, earthquake bomb being hoisted for loading into a Lancaster.

A 'Tallboy' being dropped from a Lancaster.

The *Tirpitz* lies sunk at her moorings in Altenfjorden after the raid on 15 September 1944.

Wing Commander James Brian 'Willie' Tait DSO DFC* who received a bar to his DSO with the Right Honorable Sir Archibald Sinclair, Secretary of State for Air.

Daily Express reporter, William Troughton (6th from right) who went to Duisburg twice in one day on 14 October 1944, seen here with RAF personnel on 514 Sqn including P/O John Whitwood and the crew of PD265/G (which went missing with F/O Geoffrey C. France and crew on Homberg on 21 November 1944).

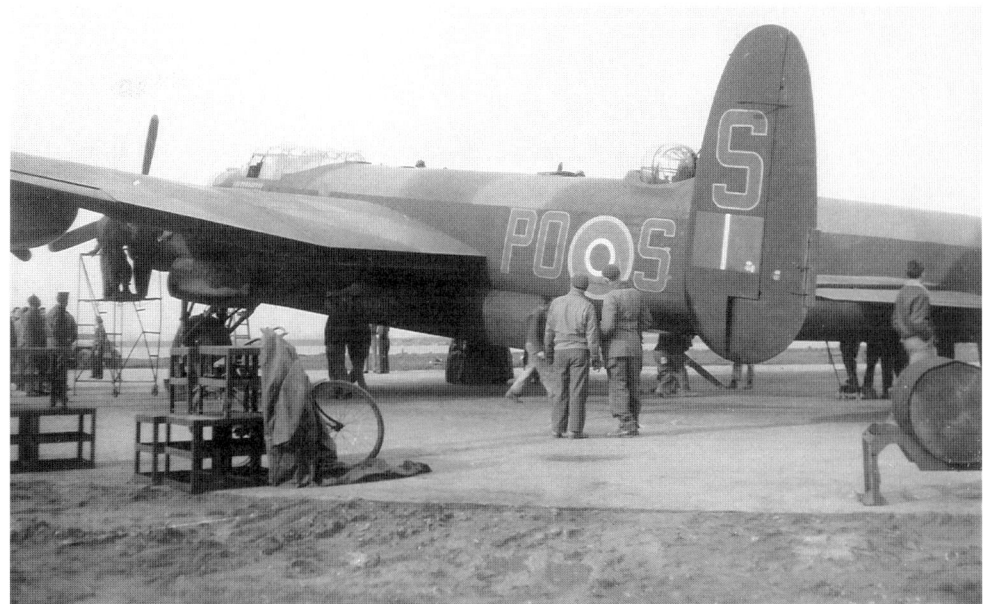

Lancaster R5868 'S for Sugar' on 467 Sqn RAAF during a stop at Seething, home to the 448th BG, US 8th Air Force. 'Sugar' completed 137 operations on 83 and 467 Sqns and flew on Operation 'Manna' at the end of the war. It is now on permanent display at the RAF Museum, Hendon.

Lancaster X KB823 built by Victory Aircraft at Malton, Ontario served as SE-E "*Lily Marlene*" on 431 Sqn RCAF and 428 Sqn RCAF (VR-U). "*Lily Marlene*" returned to Canada on 5 June 1945. (*Gordon Cross*)

Group Captain Terence John Arbuthnot removing .303 ammunition from the wreckage of Lancaster JB228, of 1668 Heavy Conversion Unit at RAF Bottesford, which crashed on landing with Flight Sergeant J. P. Jackson and his crew at Fiskerton on 10 March 1945.

Lancaster RA587 on 576 Sqn being waved off from Fiskerton at the start of the operation to Heligoland on 18 April 1945.

Lancaster III *Edith* on 218 (Gold Coast) Sqn at Chedburgh in 1945 with 'Guy' Guinane's crew. *Edith* completed 84 bombing operations on 622 and 218 Sqns and 14 food-dropping and PoW repatriation sorties to and from Holland in May 1945. Back Row: L-R: Geoff Ginn, rear gunner; 'Guy' Guinane, pilot; Jack Jarmy, navigator; Jock Lees, mid-upper gunner; Front Row L-R: Len Gillies, bomb aimer; Clarrie Ormisher, engineer; Kevin Roberts, WOP. (*Doug Gillies*)

Jubilant Dutchmen and women express their thanks to Lancasters circling during food drops over Holland in Operation 'Manna' 29 April–7 May 1945 when Bomber Command delivered a total of 6,680 tons of food. Three Lancasters were lost during the operation, two in a collision and one suffered an engine fire. Despite a German ceasefire, several aircraft returned with individual bullet holes fired by German soldiers.

'Then I dived down to 300 feet and released 2,000 loaves of bread!' — By Roland Davies

This cartoon appeared in the now defunct Sunday Dispatch on May 6, 1945.

Lancaster NN806 'M-Mike' on 576 Sqn at Fiskerton which at 1208 hours during takeoff for Rotterdam on Operation 'Manna' on 8 May 1945 swung off the runway, sending lengths of steel piping used for FIDO in all directions; the undercarriage collapsed and 'Mike' slewed round in a cloud of dust and debris and broke its back. Fortunately there was no fire and no-one was seriously injured. F/O Scott had allowed his flight engineer to attempt the takeoff and he lost control.

Lancaster B.1s on 617 Sqn on 8 May 1945. KC-B is a B.Mk.I and the other two B.Mk.I Specials modified to carry the 22,000 lb 'Grand Slam' bomb.

Five aircraft bore the name *"Aries"*, the first (PD328, a modified Lancaster I) from the Empire Air Navigation School of RAF Flying Training Command at Shawbury, headed by W/C D. C. McKinley DFC AFC was flown on record breaking trips before being modified with "Lancastrian-type" nose and tail and used on the first RAF flights to the North Pole, starting on 6 May 1945 (pictured). The second and third aircraft were Lincolns' RE364 and RE367, both named "Lincolnians".

'Dougy' Millican. (*via Clive Rowley*)

Lancaster B. III W5005 'AR-L' *LEADER* on 460 Sqn RAAF at Binbrook.

Joe laughed and found that he was not so tired and they seemed disappointed that he did not bounce them on this occasion. They ragged him as he taxied the aircraft and parked it at dispersal point.

The engines ceased their roar as Dicky and Joe switched them off and all was silent.

The battle was over for tonight. [1]

At briefing on the night of Thursday, 27/Friday, 28 April 1944 322 Lancaster crews of 1, 3 and 6 Groups were detailed to bomb the Zahnradfabrik (ZF) tank engine and gearbox factories close to the Zeppelin shed at Friedrichshafen on the shores of the Bodensee (Lake Constance) deep in Southern Germany. The Met Officer announced that there would be bright moonlight which would help to achieve better accuracy but uppermost in the minds of the crews was the knowledge that it would aid identification by enemy fighters. The disastrous raid on Nuremberg four weeks earlier had been flown in not disimilar conditions. The following day when the barmaid had opened the 'Bird In Hand' at Beck Row by the main gate to RAF Mildenhall for the lunch-time session there had barely been a half-dozen customers in and by 12.30 pm they had dwindled to three. They were all locals – not a blue uniform in sight. The local policeman came in fifteen minutes later and ordered his usual pint. He took a long draught from it, brushed his thick lips with the back of his hand and said, "Hear they took a real pasting last night.'

Now, lounging beside the log-fire in the taproom of the 'Bird' on Thursday night the police constable sipped the first of his evening's pints as the muffled sound of the Lancasters taking off for Friedrichshafen came to him. Swiftly it rode on the air currents and rose to a thundering crescendo of ear-shattering noise. He watched the girl behind the bar frantically polishing a thin pint glass. It slipped from her fingers and smashed on the floor. The barmaid, he thought, was unusually nervous tonight.

Most crews too feared the worst but the route to Friedrichshafen was carefully planned and the use of diversion and spoof raids confused

---

1. *For Those Who Wait: A Story for Bomber Crew Next-of-Kin* by Flight Lieutenant Surender Lal Berry DFC RAFVR. (Morris & Co, Dublin, 1945).

the German nightfighter controllers and the Lancasters arrived over Friedrichshafen without being intercepted. At the target, however, it was a different story. Thirty-one Bf 110 nightfighters and three Luftbeobachter (air situation observer) Ju 88s were successfully guided into the bomber stream via radio beacon 'Christa' and they wreaked havoc. Eighteen Lancasters were lost. Marking and bombing at Friedrichshafen were accurate and the bombers delivered 1,234 tons of bombs on Friedrichshafen's factories causing just under 100 acres of the town's built-up area to be destroyed including the Zahnradfabrik Friedrichshafen tank gearbox factory, whose lost output could not be made good. When the American bombing survey team investigated this raid after the war, German officials said that this was the most damaging raid on tank production of the war.

'Sammy' Berry did not fly on the operation on Friedrichshafen on the night of 27/28 April. He was confined to a sanatorium with tuberculosis and Flying Officer William Ransom, a Canadian from Toronto and a veteran of 14 operations, replaced him. Lancaster 'R-Robert' (ND781) took off at 2150 hours and proceeded to the target without incident, until that is, when at about 20,000 ft and approaching the turning point before the run into the target the first attack came from dead astern and under the tail by three Ju 88s. A second attack from dead astern upper, hit the starboard elevator and starboard inner undercarriage which burst into flames and 'R-Robert' crashed near the railway station at Sainte-Hippolyte (Haut-Rhin), 18 kilometres north of Colmar. Flight Lieutenant James Andrew Watson RCAF remained at the controls of the stricken Lancaster while his crew parachuted to safety.

The flight engineer, Sergeant Roy Clive Eames recalled: "At about 0115 we were at 17–18,000 ft approximately; I was in the nose of the aircraft carrying out 'Window' procedure when some shells came through the nose of the aircraft and realising we were being attacked, I immediately left the nose to take up my standing position beside the skipper. In this position, I saw the call light flickering. This call light is used in emergency when the intercom is unserviceable. I realized that although the rear gunner's intercom has been all right a few minutes before that it must be out of order, since we heard no report about the attackers.

'I subsequently learned that the first burst which I had encountered in the nose had also damaged the port tail plane, port aileron and rear controls. The mid-upper gunner then gave orders to corkscrew starboard and an enemy aircraft opened fire which set our starboard inner petrol tank on fire and also starboard inner engine. The skipper gave the order immediately to prepare to abandon aircraft. I feathered the starboard inner engine and also pressed its fire extinguisher to try to put the flames out. He put the nose down to keep the flames away from the aircraft and to possibly quench the flames. Realizing this was impossible, he then endeavoured to keep the aircraft straight and level. This was only achieved by keeping control column pressed tightly on his chest and I realized that our flying controls were seriously damaged.

'As part of the drill, the bomb aimer [Flight Sergeant W. S. J. McKee RCAF of London, Ontario] endeavoured to attach to the skipper's harness his parachute and succeeded after considerable difficulty. During this time the skipper asked the navigator to inform the crew of our position for the purpose of escape. The navigator told us we were approximately on the French border, 30 minutes flying time from the turning point into our target. This point was a little south of Strasbourg.

'There was at no time any suggestion of panic and this was largely due to the coolness and perfect calm of our skipper. At this time, the rear gunner was out of communication with the rest of the crew, but I heard bursts of machine gun fire from his turret. I saw that the rear of the aircraft was badly damaged and I thought that the rear gunner must have been injured. Throughout the combat, the skipper repeatedly asked for news of the rear gunner and assured us that he would look after him; I think his exact words were, "Whatever happens, he'll be OK." I told the skipper that his turret was still moving, but that was the only indication we had that he was alive. The damage caused by the second attack had damaged the call light communication.

'Sergeant Ron Hayes the mid-upper gunner was giving a commentary on the fires of the starboard wing. The captain gave us orders to bail out. I remember his words, "I'm sorry lads, but you'll have to hit the silk" and in accordance with our drill, I was the first one to leave the aircraft at approximately 12,000 ft. I acknowledged the captain's order as I left, and that was the last time I saw him."[2]

---

2. Eames was followed by McKee and Pilot Officer W.H. Russell, wireless operator and Ransom, who saw Watson still trying to control the aircraft while the rest of the crew abandoned it.

'The aircraft was approaching our turning point before the run into the target when it was attacked from dead astern under' Hayes (19) recalled later. 'The attack was a complete surprise; there was no moon, just complete darkness. The aircraft was equipped with H2S radar equipment which transmits pulses and the crew and Intelligence was not aware at the time that the Germans were able to home in on the signal. The first attack came from dead astern and under the tail, by three Junkers 88 night fighters.

'As the aircraft was attacked, from the rear thuds were heard at the rear and flashes and the port elevator was badly buckled. The rear gunner [Flight Sergeant Murdock Daniel MacKinnon RCAF a Cape Breton native living in Somerville, Massachusetts, when he signed up] was out of communications and could not direct the pilot on evasive manoeuvres, so I took control of directing the pilot with evasive direction. From the bursts of fire, we were under attack by at least two attacking aircraft and we could not see them, so Jimmy decided to keep the aircraft on course, rather than attempting to dive away from the attacking aircraft, which was what the attackers would be expecting. A second attack from dead astern upper hit the starboard elevator and starboard inner undercarriage, which burst into flames. As the attacking aircraft was coming in closing in from the starboard quarter level and at about 350 yards Jimmy was directed to 'corkscrew' to starboard. His immediate evasive action, even with the badly buckled port elevator showed that he had the aircraft under full control. Jimmy's response to evasive direction was magnificent, but the aircraft was hit about the starboard inner engine and a second later this portion of the wing burst into flames. The first impression was that the starboard inner engine was on fire but from dialogue between crew members in the cockpit, it was determined that the fire extinguisher system had been activated. Jimmy was in full control of the aircraft, but the fire did not die out as was hoped for by the crew. The danger of flames was increasing all the time and he side slipped the aircraft to keep them away as much as possible, as the aircraft kept losing height at the same time.

'The flames were causing the seam aft of the starboard inner engine to melt and Jimmy was informed of this, who then ordered everyone to collect their parachutes. The aircraft continued to lose height and the flames had enveloped most of the wing and half of the seam had melted.

Jimmy was informed of this and he ordered everyone to bail out. I plugged into the intercom system and informed him that I was bailing out and that the rear gunner was still in his turret and to let him know we were getting out. Jimmy's last words to me were 'Yes, OK, but hurry, we're at 4,500 ft, if he's not hit he might make it. So long Ron, good luck.'

'I opened the bulkhead door leading to the rear turret and saw Murdock MacKinnon turn his head towards me, I patted my parachute to indicate that we were bailing out and he understood. The aircraft was now at about 4,000 ft when I bailed out. Jimmy had the aircraft under perfect control. It was still losing height in a sinking fashion and the flames had enveloped the fuselage alongside the burning wing.

'I landed hard in an open field, landing on my right foot and fell or was pulled onto my right side; and dragged some distance by my open parachute canopy until it collapsed. The action with the German fighter aircraft, the difficulty in evacuating our aircraft and the bail-out and hard landing in the dark were very stressful experiences and the right side of my body and lower back was aching. I experienced some dizziness, so I rested for a few hours where I had landed, out in the open. With daylight approaching, I stood up to walk in search of a hiding place, for a wood or an isolated barn, but experienced disabling pain and only managed to make it to a nearby ditch, where I was discovered, by a man, an Alsatian and taken to Guermars at about 0100 hours on 28 April. At this village I was interviewed by a young girl who could speak a little English and I was then taken to the village hall.

'Here I met a French schoolmistress, Mademoiselle Louise Strohl, who gave me tea, biscuits and tobacco and then she told me that Flight Lieutenant Watson had been found dead at the controls of the aircraft. She went to some length in describing him, even saying he was a Canadian and that he had two stripes on his epaulettes. This lady was sympathetic and wanted to cheer me up and make me feel at home, even though she could not help me escape. The village hall had become crowded with the local inhabitants who might have helped me escape if it was not for their fears of the Gestapo. From here, I was taken by two Luftwaffe Intelligence Officers to Colmar, where I was interrogated. After the usual questions, I was asked if I could help them in identifying the belongings of a dead pilot. The items were those of Flight Lieutenant Watson in an envelope, consisting of his identification bracelet and a ring. I knew that the ring had been given to Jimmy by his father. The Germans said that they had taken the articles from a pilot who was found dead in his seat of a Lancaster. I said nothing to them for fear that it might be the beginning of a long interrogation and I also knew that the identity bracelet was sufficient.

"At Colmar, I thought about it and formed the opinion that the pilot had died in an attempt to save the rear gunner and had attempted to execute a crash landing. I saw the wireless operator, Pilot Officer Russell, Ransom and Eames but they did not speak to each other as the Germans might be listening. Eames and I were taken to Stalag Luft 6, while Ransom and Russell, being officers, were separated and there was no opportunity to talk in quiet. On way to Stalag Luft VI, I learned from Eames that he had seen MacKinnon the rear gunner arrive the day before and had received quite a shock because both of us thought that he had also been killed in the aircraft.'

James Watson is buried at Choloy War Cemetery at Meurthe-et-Moselle France. Roy Eames said: 'It is quite clear that he sacrificed his life knowingly and willingly to ensure the safety of his crew. His most courageous act, his great and noble sacrifice in the face of the enemy was beyond the highest ideals of his duty and merits the highest possible award for gallantry and for valour in the face of the enemy.'[3]

In a letter to his father (28 April 1944) Wing Commander Ian Clifford Kirby Swales DFC DFM Commanding Officer, 622 Squadron wrote, in part: 'Your son was one of our most experienced pilots, and had successfully completed sixteen operational flights. He was admired by all who knew him, and will be much missed by his many friends in the squadron.'

That morning Sammy Berry awoke with a strange feeling. Unlike most of his dreams, he remembered this one in every detail. It seemed too real to him to be a dream at all. He rose and drew aside the blackout curtains. He looked out of the open window and saw the aerodrome, with the aircraft still and silent. The sun shone on them from the South East.

'What a lovely day,' he thought and noticed the sky's azure blue, which faded into a soft grey-white as his eyes lowered to the horizon. The sky

---

3. In 1946 and 1947, five members of the crew put forward recommendations for the Victoria Cross to be awarded to Flight Lieutenant James Watson but he received a posthumous Mention in Dispatches only.

was clear but for one cumulus cloud. Its dazzling beauty held his attention. He studied it closely and saw the cloud was slowly growing bigger. The grey shadows in this otherwise rich white cloud were moving and forming different shapes. He stood fascinated by it until he saw the shadows form into a face. It was a fresh looking face, inconspicuous, but handsome. He almost fell out of the window as he recognised it. 'What did it mean?' he asked himself. He felt he knew now that his skipper had been killed and realisation came upon him. He understood his own unquenchable thirst. It was not himself, but his skipper's spirit and strength, which was driving him on and on. He knew his dream could never come true. He looked at the cloud again and saw another face. He saw his mother just as he had always remembered her – she was smiling at him. Suddenly, the vision was gone.

Sammy was in a reflective mood. He suddenly thought he would like to refer to his old diary and see how he felt this night, last year. He stood on a chair and, from the top of the wardrobe, brought down an old suitcase. It was a suitcase which he had left there when he went to hospital. It was full of old mementos, which he liked to have with him but which were of little practical use to him. He blew the coat of dust off and then inserted a key to unlock the case. To his surprise the key would not turn and he found that it was not locked. He opened it and found everything as his skipper had left it, but there was an addition. It was a book. He picked it up and looked at it. To his amazement he discovered it was his skipper's memoirs. He read it and lived through every operation he had done without him. Then he came to the last. His skipper had written it just before take-off on that last trip and he wondered how it should come to be in his suitcase, which had been left locked. 'What did his skipper mean?' he thought. 'It seems almost as if he knew something was going to happen.'

'I write this before I go to briefing, which will be in a few minutes. I am filled with a queer feeling, which I can't define. I wonder if I am just bomb happy, for this feeling tells me that it is a tougher target than ever before. I have but three more operations to do and yet, with all this experience behind me, I feel like a beginner. I have no idea what to expect. I only know that if I live, I will always remember these days as the greatest of my life. The days of hilarity, mingled with despair. The days of supreme vitality with those of exhaustion, the days of complete happiness and tragedy and the days when there was only

the present. Day by day, each hour, each minute was strange, real and alive. So much happens in such a short time that it seems a lifetime; a lifetime of extremes when we appeared so old and yet so young. So sad and yet so hysterically happy. Yet, with it all, I have a happiness which has never been surpassed, as I live from day to day and know that I am still alive. A happiness which is complete and satisfying. The happiness which is inspired by the love of men and God, in a noble sacrifice that the world might resume its natural, peaceful serenity.'

Sammy washed and dressed. He looked at his skipper's memoirs and knew the message he had been trying to tell. He sat down and continued his skipper's memoirs in pencil. He wrote: *'From the facts I have learned about this operation, coupled with the feelings in me of which I feel sure. I dare to continue this flight on skipper's behalf.*

'Our target was Friedrichshafen on Lake Constance. This lake has Switzerland on one side and Germany on the other. We had little trouble over Southern France – it was pretty much the same as usual. The odd bit of flak and searchlights: but, when we came to Germany, tough opposition began. We were attacked three times on our way to the target. The sky seemed to be infested with Jerry night-fighters. We suffered little damage ourselves and Mickey and Tommy are sure they damaged one between them. We arrived at the target on time. Good old Steve, prompt as ever. We bombed OK and Johnny was pretty sure he was on the button. I could not help thinking, on our way back, that's only two more to do now. But I remembered that this one had still to be finished.

'Then it came. 'Fighter, fighter' from Mickey and I turned, but at that very moment Tommy said 'Fighter, fighter,' the other side.

'We had been warned of the German's new tactics of simultaneous attacks and while I pushed the stick forward into a dive, I had a terrible feeling of knowing that Tommy and Mickey were firing at different 'planes. We lost one and Tommy said he had hit it, but lost sight of it as it went down. The other continued firing on our tail and Tommy assisted Mickey. Suddenly, there was a swishing past my head – it was like a flash of lightning – and I realised, as I saw a line of holes, how narrowly 'T-Tommy' had missed being hit. We continued firing and I felt this was the longest battle we had had to date. Then the port outer engine was hit and I had to feather it. This loss of power only served to add to our predicament. Then this fighter came to the other side and got our starboard outer engine, but all was suddenly silenced as Mickey said: 'I've got him at last 'Skip'. I can see him go down – he's on fire.' Thank heaven for that, 'Mick.' We continued on our route, but, with two engines, had to go slower

than we were briefed. We knew this was further endangering our position as time went on. The mass of aircraft were steadily leaving us behind and soon we would become victim to all the German could give.

'I cannot say I felt afraid – I don't know what it was. It was like knowing something was going to happen and not being able to do anything about it. We might be lucky – on the other hand, we might not – but I had a queer feeling, as we slowly made our way back, expecting the worst.

'For an hour all was quiet and Steve said: 'I calculate the stream is now fifty miles ahead of us, but we're on the French frontier now; so perhaps things won't be quite so hot.'

'We dodged small areas of flak over France and then it started. Not one fighter, but three. We knew it was now a fight to the finish. These Huns were after Iron Crosses.

'The battle continued and I was exhausted, but I knew I must not weaken. It was a terrible moment when the port wing was on fire. I side-slipped to keep the flames away, but the fire showed no sign of diminishing. I knew now that it was only a matter of time before tanks exploded. We went down in a gradual descent.

'I gave the order: 'Chutes on – prepare to jump.'

'They all acknowledged. As we went down I kept the flames away from the wing as best as I could.

'Just as I was about to say 'OK fellows, this is it, out you go – jump!' another swish of lightning came past me and I was thrown forward by a terrific jolt on to the joy stick. I saw the stream of holes around me and I realised I had been hit, although I felt no pain.

'As I heaved myself off the stick and pulled the aircraft out of the dive, I felt the hot blood rolling down my back. Then the pain started and it grew stronger and stronger – it was paralysing me. Suddenly, I forgot myself completely. I looked at the altimeter. We were at eight thousand ft, but we were losing height fast.

'The crew did not know I had been hit and I decided not to tell them, as I gave the order to jump.

'So long 'Skip',' said Johnny.

'What more can I say 'Skip'?' said Dicky.

'I'll be seeing yer' rasped Mickey.

'Well done Skipper' said Steve.

'My wife loves you' said Harry.

'Come on 'Skip', I'm the last' said Tommy.

'I smiled to myself as they left with such remarks. Even to the last, they knew what to say to fill me with strength. I was glad I had not told them I had been hit, for I know they would have thought about it too much afterwards. It is strange, as I know they are safe, I can feel no pain – nothing except an exaltation of peacefulness I have never felt before. I know I have only a few minutes to live. I cannot move myself – my body is completely numb. As I think here, I feel outside myself, I am not in that body which is hurtling down towards the earth. I belong up here and here I shall stay.

'My heart is heavy for only one thing: I shall never see my dear family again. It is the only thing I really want, as I have a message for them. I want to tell them that I love them and that I shall always be near them. I want to say thanks for a happy life.

'It grieves me, as I realise how they will suffer. I know they will wait day after day and hope. Then, when I have been presumed killed, they will suffer more. They will continue to wonder and still hope. They will be filled with dreams, visions and prayers. In this, my last minute, I think of my people and the thousands upon thousands like them. I know then that the greatest suffering is not on the battlefields: it is in the homes of those left behind. I turn to God and pray for them.

'O God, help those who wait and suffer anxiety that they can endure, hope, and pray.

'And for my people and those whose dreams cannot come true, for those who hope in vain! Give comfort and courage, O God that their loved ones died in the cause of righteousness. Help them in their hopes, their dreams. Keep them from despair and disillusionment. Hear me that my prayers be answered, for those who wait...'[4]

---

4. Adapted from *For Those Who Wait: A Story for Bomber Crew Next-of-Kin* by Flight Lieutenant Surender Lal Berry DFC RAFVR. (Morris & CO, Dublin, 1944). 'Sammy' later joined the crew of F/Sgt William Kenneth Thomas (later DFC). On 8 November 1944, though severely wounded, Berry performed an act of conspicuous gallantry in navigating their badly-battered Lancaster to the safety of the English coast. He was awarded an immediate DFC.

Chapter 8

Orphan Airman

Geoff Reynolds

O*n the night of 31 May/1 June 1944, a Messerschmitt Bf 110G-4 with its deadly electronic wizardry and heavy firepower flown by 29-year-old Hauptmann Fritz Söthe of Stab.ll/NJG4 at Coulommiers climbed at maximum rate to an operational height of 5300 metres to the west of and slightly below the bomber stream of 125 Lancasters and 86 Halifaxes on their 325 nautical mile flight to the rail yards at Trappes. Söthe, born in Lohfeld/Minden* on 20 September 1914, *was seeking his tenth Abschüsse* [victory], *having by now destroyed nine 'Viermots' (four-engined bombers) while on Nachtjagd duty. As he continued on an almost parallel course to the east of Paris his bordfunker* [radio/radar-operator], *20-year-old Unteroffizier Wilhelm Brönies from Hamm/Westfalen looked for contacts on the Lichtenstein radar set and the 21-year-old bordshütze* [engineer-gunner], *Unteroffizier Heinz Enke of Rositz/Altenburg, scanned the night sky. Eagerly but patiently, Söthe awaited the code word from the Jägerleitoffizier* [JLO or GCI-controller] *in his 'Battle Opera House' that would send him scurrying into action in one of the 'Himmelbett Räume'* ['four poster bed' boxes] *each one of them a theoretical spot in the sky, in which one to three fighters orbited a radio beacon waiting for bombers to appear.*

The JLO took some time to bring Söthe into a position to make contact with the bomber stream because of the countermeasures employed by the RAF, announcing monotonously at regular intervals, 'No Kuriere in sight' and Söthe had to continue orbiting. Then, suddenly, as if by magic, 'Have Kurier for you, 'Kirchturm 10' [1000 metres], *course 300°, Kurier flying from two to 11' sounded in the Bordfunker's earphones.*

Söthe began zigzagging around the sky in the hope of finding the contact. Then, by a stroke of good fortune, at Meru, 40 kilometres north-west of Paris, he flew through the turbulence created by one of the bombers and Brönies at once picked up its signal on his scope. Söthe had already decided to use his nose armament against the 4-mot. It was 'D-Dog' on 622 Squadron at RAF Mildenhall piloted by Flight Lieutenant Francis Reginald 'Frank' Randall.

Attacking using the conventional technique 'von unten und hinten' [from below and behind] *to take advantage of the bomber's blind spot, Söthe achieved complete surprise and it is doubtful that the mid-upper gunner and Flight Lieutenant Leslie Frederick 'Peter' Berry* DFC MiD *the 36-year-old tail-gunner ever knew what hit them.*

At 100 metres Söthe pressed the gun button on the stick and was startled at the rattle of the cannon. He stayed behind the great night bomber, deliberately and accurately firing short bursts of brightly coloured tracer that peppered the Lancaster's fuselage, wing and the fuel tanks in the wings before diving away. Randall did not even have time to Corkscrew. The fusillade of cannon shells and machine gun fire ruptured the fuel tanks starting a huge fire near the port wing root, which quickly spread. Doomed, 'D-Dog' fell away to port in a flaming death dive and the Messerschmitt was thrown about vigorously by the force of the resulting explosion. There was no doubt about the fate of this bomber.

In minutes LM121, C for Charlie, Söthe's second victim of the night, beckoned. It too was from Mildenhall, one of four Lancasters shot down on the raid on Trappes, three of them from the Suffolk station, a 10% loss rate for the base.

The time was now 0154.[1]

Over northern France the throb of the four mighty Rolls-Royce Merlin engines was transmitted through the airframe of 'C for Charlie' on XV Squadron at RAF Mildenhall. The crew skippered by Pilot Officer Peter Charles Lewis D'Ombrain RAAF were on their eighth operation. During the afternoon of 31 May they had been one of the crews on *'Oxford's Own'* together with those of 622 Squadron that were briefed for the raid on the railway yards at Trappes in the south-western suburbs of Paris by 125 Lancasters, 86 Halifaxes and 8 Mosquitoes. The 'rookie' crew was part of a second wave that would hopefully be bombing a well-marked and clearly defined target area. It was to be a busy night for the 800 aircraft of Bomber Command heading for Occupied France to bomb not only Trappes but railway targets at Saumur and Tergnier, a coastal gun battery at Maisy, a radio jammer

---

1. Berry evaded capture and returned to England with the assistance of the Resistance. The six other crewmembers were killed. *Bomber Command: Reflections of War – Battles With the Nachtjagd, 30/31 March-September 1944, Vol.4* by Martin W. Bowman.

at Mont Couple and a transmitter at Au Fèvre. All were meticulously planned and if well orchestrated could add extra confusion to those waiting to pounce on them from above or attack them from below.

The first members of Peter D'Ombrain's crew had come together at 11 OTU at Westcott and Oakley in October 1943. Born at Shapperton, Victoria on 30 September 1922, the son of Ray Fenley and Ethel Mary D'Ombrain of Mildura, Victoria, Western Australia, where his parents ran the Grand Hotel. Peter was educated at Melbourne Grammar School and he enlisted in 1940. Training at Somers, Benalla and Mallala in South Australia soon followed. Having grown up in and around the hotel trade back home in Mildura, Victoria he found it easy to mix with most people. Not quite 22 years-old, he did however look much older, something he put down to the hot Australian sun, the beers and the hard studies and training needed to reach his lofty position. Like many Australians serving in the Royal Air Force he had a strong, determined disposition. He threw himself enthusiastically into everything that he and his crew did. His sometimes reckless and boisterous bouts on the ground, especially if his national pride was offended and the beer was flowing freely, had never shown itself in the air. From the moment that the crew came together at the OTU he had begun to lead them with an assertiveness and authority that belied his young age and his all too recent student background.

Flight Sergeant D'Ombrain's crew had been posted to Mildenhall, to XV Squadron, on 1 May 1944. The skipper's recent promotion to pilot officer had been popular with the crew and one they had celebrated heartily. Known to the crew as 'Skipper' or 'The Boss' (the latter given him by their young flight engineer), his outlook was not going to change with his promotion. As the pilot, no matter what his rank, he was always the Skipper and in charge of all the crew on the aircraft. Even if Wing Commander William David Gordon-Watkins DSO DFC DFM, XV Squadron's Commanding Officer had been aboard, Peter would still be in charge.[2]

---

2. Gordon-Watkins was shot down on 16 November 1944 whilst piloting the lead bomber on Heinsberg. He was the only member of the crew to survive, was captured and held as a prisoner of war in Stalag Luft I. He had completed over 50 operations and had previously served on 149 Squadron.

In February 1944 when he and his crew converted from the Wellington to the Lancaster at the Heavy Conversion Unit, two more members joined them. Sergeant Leslie Arnold Hadder the incoming air gunner was born in Bromley on 30 October 1924. His first work experience, from the age of 14, had been as a motor mechanic's mate. Then the Hadder family moved to Thorpe-next-Norwich and on 31 March 1942 Leslie volunteered for the RAF.

The other new recruit was Sergeant Leonard Thomas Gearing the 19-year-old flight engineer. This position needed to be filled because it had not been a requirement on the Wellington. Before volunteering for the RAF he had worked for the Ford Motor Company in Dagenham, which was then a quiet Essex village. A very talented footballer, he had played for Dagenham with Alf Ramsey.[3] Gearing would hang around the dispersal area before raids talking to the ground crews. From the information they were able to give him on the fuel and bomb loads, he could begin to calculate the expected distance to the target. There was not much about the great aircraft that Len did not know something about, except the radios and much of the electrical equipment.

Gearing was one of only two Englishmen on the crew. The other, Sergeant Raymond Geoffrey Norris, the 21-year-old wireless operator, known as Geoff, was the only crew member who had the patience to deal with the radio equipment's little idiosyncrasies. None of the other crewmembers understood these and he would be the one to sign off the Form 700 to state that all things electrical were satisfactory. He too got very friendly with the ground crews and could often extract the necessary information to pass on to Len. Armed with this information and any gleaned from other crews the crew could begin surmising about the possible destination for the coming night's operation. If the fuel load was 1,150 gallons it was to be a short trip, 2,154 gallons -full tanks – and they were in for a long one. 'C-Charlie' had 1,857 gallons when the crew had set off; enough for at least six and a half hours in the air if required.

On the night of 31 May D'Ombrain took 'C for Charlie' off from Mildenhall at 2357 hours (the fourth aircraft away) with a 10,000lb load

---
3. Ramsey (born, 22 January 1920 and raised in Dagenham, went on to become a famous Spurs and England full back and he later managed Ipswich in the Championship winning season (1961–62) and led England to World Cup glory at Wembley Stadium in 1966 when they beat West Germany 4–3. Sir Alf Ramsey died on 28 April 1999.

and the necessary fuel to reach the target and return safely. The first half of the outward flight was uneventful. Climbing steadily away from Mildenhall they had formed up with the main bomber stream, taking up position towards the rear of the 2nd wave for the attack on Trappes. Occasionally, they felt the turbulence of other mighty Lancasters. It was slightly comforting to know that they were all heading for the same goal and that one didn't have to run the gauntlet alone.

Just over an hour and a half after takeoff the French coast at Fécamp drifted slowly by beneath. 'C for Charlie' had entered the 'lions' lair'. D'Ombrain banked gently to port onto a new heading and the Lancaster droned steadily on across the dark expanse of occupied Europe. Soon the sky would be getting lighter in the east. From their lofty position, of almost 17,000 ft, the crew would be among the first to witness the sunrise over this part of war-torn Europe. Threateningly, but far off in the distance, lightning flashed from cloud to cloud and, unheard by the crews, a distant thunder rumbled across the war ravaged land.

Geoff Norris needed to relieve himself. He unplugged his earphones and disconnected his oxygen mask to make the difficult and arduous journey to the rear of the aircraft, to the little Elsan toilet that was situated just in front of the tail turret and tail-plane strut. It meant leaving the warm and normally comforting surroundings of his little radio cabin and his seat in front of the dials on his T1154 /R1155 transmitter and receiver to make the journey over the bomb bay and the main spar in his flying kit. The excursion was doubly difficult because he had always made it a rule to clip on his parachute whenever he moved about the aircraft in flight. Sitting on it at his radios, he found it a comfort to know exactly where it was in case he ever needed it, which he fervently hoped he never would. Without the benefit of oxygen, which was always used above 10,000 ft, the 80-ft long return journey was like an eight mile run but he hoped that the visit would settle his stomach; he wasn't altogether sure it was something he had eaten or drank that was causing the problem or his nerves taking over.

Tonight, more than any other since beginning operations, he had a deep sense of foreboding. He could not put his finger on it nor had he been able to share it with the crew. He was sure they all felt frightened to some extent, but this was something different to the fear they all faced each and every trip. It had come on towards the end of his last leave, at home on the Isle of Wight. He had tried to share his thoughts with some of his friends in Carisbrooke where he had once sung in the choir at the parish church but he had found it difficult to put them into words. The change in his general demeanour had not gone unnoticed. Gone were the boyish glint of fun in his eyes and the laughter lines of his young face had been replaced by a drawn look. Jean Mullins, his elder cousin, remembered him as having a 'certain presence; there was 'definitely an aura about him. Most of all he had charisma.'

Close to a rather favourite apple tree in the orchard at the back of her house at Fairlee on the road to Ryde, Geoff had suddenly turned to her and said: 'D'you know Jean, I shall never return to the Island again.' Without looking at him she replied: 'Oh Geoff, don't be so silly. Stop talking like that. How do you know?' 'It's true' he continued rather matter-of-factly. 'The odds are so stacked against me and the lads. I know I shall never come back again.' As Jean had turned to face him, in a calm voice he said, 'You mustn't worry about me, I just know it's going to happen and it will be alright you know.' With tears welling up in her eyes, Jean and Geoff turned around and walked on through the orchard in silence.

Geoff knew that those who had not been over enemy territory could not guess what it was like. The last thing he wanted to do was give even the merest hint that he was frightened. This had especially applied to his family. His late father would have known; having spent the Great War in the trenches of France. He had been diagnosed as having incurable cancer and died in April 1932 aged only 39. His wife's health declined, until after gamely struggling on, she eventually succumbed on 11 October 1933, aged just 46. Geoff had been born in the early hours of 6 August 1922, just 16 days before his cousin Eric, whose parents raised Geoff at No.2 Ashengrove Cottages, Gunville; a small settlement on the Isle of Wight. He had worked at Fisk and Fisher at St. Cross Mill in Newport before volunteering for the RAF in the hot summer of 1941 at age 18. Muriel Gill, his cousin from Bournemouth, recalled many years later that Geoff told her he had done so to avoid being conscripted into any other arm of the forces.

As was usual, on the morning that Geoff was to due to return to Mildenhall he had awakened Eric early to say 'cheerio' but before his cousin left for work Geoff hugged and kissed him. Although they had become blood brothers, this level of closeness had never before been evident. Nobody took much notice of this or his somewhat out of character, morbid predictions, but all was to become painfully clear at the end of the month.

Geoff settled back down in the warm, familiar surroundings of his radio room and glanced at his watch. From the dim glow of the many dials he could see that it was 0141 hours. 'C for Charlie' still smelled very new, the odours of metal and fabric, paint and oil, of cordite and fuel and all those peculiar smells that identified this aircraft to her crew. Over the course of operations this would slowly change and would become 'Charlie's' unmistakable smell. Being factory fresh and having only completed one operation so far, the whole crew was hoping that 'Charlie' would remain 'their' aircraft and bring them luck. When they got home one of the ground crew could stencil a second bomb symbol beneath the cockpit windows. 'Charlie' and the crew still had a long way to go to match some of the others on the squadron.

Despite his feelings of doom and gloom, Geoff gained some comfort in the knowledge that the crew had really come together as a fighting team. Now that each member of the crew knew his own individual jobs far better, the crew would happily follow where the skipper led. Each crew member had faith in each other's abilities. As a tight knit team they alone had the tools for their survival each time they took off on an operation. No matter how many other crews were operating on any given 'op', every crewmember functioned as a separate entity. Without team spirit, each could have felt very much alone in the fight, as though each enemy bullet and shell was directed at him personally. They needed each other, for only working together, smoothly and efficiently, could they all get through safely.

Although a seat was provided for the flight engineer it was impossible to see the instrument panel clearly and Len Gearing chose to stand for most of the operation, except during take-off and landing. Shy and quiet by aircrew standards, Gearing was methodical and could always be counted on to have the up-to-date fuel situation. He knew his four favourite mighty Merlin's were the key to the Lancaster's wonderful load carrying and flying abilities. He and 'The Boss', as he had taken to calling Peter D'Ombrain, worked together to ensure that they got the most from these engines. They carefully balanced the need to conserve as much fuel as possible, should it be needed for evasive tactics, the desire to get to the target on time and then home again as quickly as possible. During takeoffs and landings Len would sit and assist 'The Boss' by operating the throttles and flaps. He would call out the airspeeds and raise the flaps and undercarriage so that his pilot could use both hands on the control column and all his strength to pull back and get the heavy bomber into the air.

It was slightly unsettling that since joining the squadron the crew had not had a regular mid-upper gunner. Leslie Hadder never got to fly on D'Ombrain's crew. He flew three trips with other crews but had to convalesce at the Hoylake Remedial Centre in Liverpool following three months in and out of Ely RAF Hospital after receiving several injuries in crash landings.[4] Three different gunners from a 'pool' occupied the mid-upper position in turn on Peter D'Ombrain's crew but had not had time to settle in. For the raid on Trappes, 21-year-old Australian Flight Sergeant Stanley Arthur Nystrom, usually known as 'Sam' for some obscure reason, was occupying the mid-upper turret. He had flown with them two nights before and once, earlier in the month. Born at Kingaroy on 17 February 1923, 'Sam' was the son of George Arthur and Ada Ethel Nystrom of Annerley, Queensland. Enlisting in the RAAF on 31 January 1942 he had been on XV Squadron for some time. Trappes would be the 19th operation of his tour. The crew had taken to him and he was pleased to be flying with them again. Perhaps the crew would be able to persuade him to become their regular seventh member, at least until his tour was finished. Like many of the Australians he played and worked hard and was dedicated to his duty. Although he had a cavalier attitude to life in general, he had the ability to totally concentrate on the job in hand and he made a valuable addition to the crew. He would certainly not be brooding over any impending disaster and was probably thinking about the beer and girls who would be attending the dance tomorrow night!

---

4. Flight Sergeant Hadder was killed on the night of 3/4 March 1945 on Fortress II HB815 BU-J on 214 Squadron which was shot down by a German intruder in the circuit at RAF Oulton in Norfolk at 1840 hours on return from 'Window' duties on bomber support. (See *Confounding the Reich* (Pen & Sword, 2004); *Voices In Flight: The Night Air War* (Pen & Sword, 2015) and *German Night Fighters Versus Bomber Command 1943–1945* (Pen & Sword, 2016) by Martin W. Bowman. Hadder is buried in the Thorpe-Next-Norwich (St. Andrew) Church Cemetery.

Geoff thought that perhaps his fears were unfounded. After all they had survived two engine failures in early training. Both times they owed their lives to D'Ombrain's skill as a pilot and the sturdiness of the aircraft. They had come through the first five operations when most 'rookie' crews were prone to perish. Trappes was only their eighth operation but they had twice been to 'Happy Valley'. Germany's major industrial area in the very heart of the Ruhr and one of the most heavily defended areas in Hitler's Third Reich. They had returned when more experienced crews had not. Geoff guessed that, statistically, it was impossible to complete a tour of 30 operations; it didn't take a mathematician to work that out! Yet crews were surviving. Colonial flyers from all parts of the Empire returned to their homes on completion of their tour. Yet many stayed to fight again or teach at training units, passing on to new crews the knowledge that had seen them safely through. If they maintained the disciplines they had been taught, then surely 'C for Charlie's crew would get through, wouldn't they? Why didn't these thoughts remove the feeling of gloom that Geoff felt?

Flight Sergeant Laurence Seymour Jamieson, the 26-year-old bomb aimer carefully scanned the skies ahead and on each quarter, both above and below, watching for any landmarks that he could point out to the navigator and any sign of an enemy fighter. He was by far the oldest member of the crew. Unlike the skipper, he looked much younger but the others affectionately referred to him as 'Pop'. Born at Easter Skeld in the Shetland Islands on Christmas Day 1917, Jamieson had emigrated to New Zealand with his parents when he was 8 years old. He was more quietly spoken than the Australians on the crew and he retained a faint hint of the Celtic accent. 'Pop' could usually be relied upon for some fatherly type advice on most subjects. He was just as likely to be found drinking in the local pubs or kicking or throwing a ball about on the airfield. He had always enjoyed a game of rugby back home and competed with vigour. These ball games and drinking sessions were always a source of friendly rivalry between the various nationalities that made up the crew and indeed the whole squadron. His previous life as a mechanic for the Hope Gibbons Cycle Company meant that he was much in demand by many on the station where bicycles were the major mode of transport.

For much of the flight 'Pop' would occupy the position of front gunner. Few fighter attacks came from this quarter but he was able to keep a sharp lookout at all times, even for friendly aircraft that might converge on their course. Nearer the target he would lay in the nose of the aircraft as bomb-aimer, returning to his turret only after that important task had been completed. He would look forward and downwards through the bombsight, right into the holocaust of searchlights, gunfire and explosions while the aircraft was flown as straight and level as possible, a sitting duck for all the gunners on earth to shoot at. His precise and deliberate directions to the pilot during these tense moments would ensure that they passed directly over their allocated aiming point. He would then release the bombs at the precise moment to send them screaming down into the very heart of the target.

With the bomb load gone the engine note would seem to change and the Lancaster would become a more graceful flying machine. In the hands of her capable pilot who had learnt to get the best out of her, she would soon be flying like a bird. Once through the searchlights and flak and away from the target they could turn and set course for Mildenhall, where, following debriefing by the station Intelligence Officer, they could grab something hot to eat and drink before retiring to their beds; hopefully before an awakening East Anglia was able to interrupt their well deserved rest. To Geoff, all this seemed a long way off as they headed east by southeast towards their next turning point.

Geoff could look back and make out 'Sam's lower body protruding from the bottom of the mid-upper turret, which in the gloom he could just see rotating slowly in each direction. Geoff knew 'Sam' was scanning the unfriendly skies for the enemy aircraft that must be out there somewhere. 'Sam' came from Annerley, a suburb of Brisbane in Queensland, the same city as the third Australian member of the crew, 21-year-old navigator, Flight Sergeant Arthur Stephen Long, who hailed from the Hawthorne area. Maybe 'Steve' would have a chat with 'Sam' and get him to join the crew for the remainder of his tour. Geoff knew instinctively that they would all get along just fine.

Geoff and 'Steve' had begun a close friendship at OTU Westcott. Although from different backgrounds, brought up half a world apart, they had a passionate love of nature and the environment. When off duty they enjoyed long walks in the local countryside. 'Steve' discovered that the wildlife and landscape was so different from that of Australia. When the base bar and the 'Bird in Hand', were running short of beer, as they so often did, the pair of them would spend many an interesting and relaxing time in the lanes and fields, sometimes venturing further afield using borrowed bicycles from the many on the base. Often they were lucky enough to find a source of good local ale in a remote country pub, away from the tensions of base life, where they would talk about Australia and the Isle of Wight, quench their thirsts and enjoy a chat with the old regulars. Each was eager to learn about the other's past and how different their lives had been. 'Steve' never complained about the English climate, for him it was just another part of nature that made this little island such a beautiful and interesting place. It was 'Steve' who had managed to acquire an Australian issue, pure sheepskin flight vest that he had given to Geoff in the depths of the previous winter. Geoff had left it at home on his last leave. Now that the better weather was here he felt he would have no need of it through the summer months. How naive and wrong he'd been. Although quite bulky under his flying clothes, it would have been a bonus back at the Elsan; the RAF issue clothing never did seem as warm.

'Steve' reminded Peter D'Ombrain of the next change of course. They would be turning due south in six minutes' time. Hopefully, somebody would spot the small lake that was their turning point, but ground mist was blotting out many of the features and he would have to rely on his timings. This turn would leave Paris on their port side as the second wave ran in for a timed run to the target. Soon, coloured TIs dropped by the Path Finders were raining down, followed by explosions as the first bombs hit

D'Ombrain acknowledged the change of heading that 'Steve' had given him and he checked his watch. It was now 0146. As skipper he was always courteous but short in his replies, often not repeating the command to save time. He did not allow any useless chatter over the intercom, a rule that the normally vociferous and sometimes coarse Australians adhered to. They all knew how important the next call might be to their very survival. Any banter might just mean that their pilot missed the first warning call from one of his gunners and a split second lost in execution of the evading manoeuvre could mean the difference between life and death.

Flight Sergeant Frank Bruce Reid, the 21-year-old, pipe smoking rear gunner, was born at Berwick, a suburb of Melbourne to the Reverend Frank Reid and his wife Ethel on 4 December 1922. On enlistment Frank was living at North Geelong in Victoria in south-eastern Australia. Sitting cramped into the tiny, freezing confines of the tail turret, sixty feet behind Skip, he kept his lonely vigil over the night skies. His was certainly the most unenviable position in the whole aircraft. Commonly known as 'Tail End Charlie', none of the crew would willingly swap seats with him. This cheeky Aussie, always first to part company with the ground, as the aircraft strove to attain flying speed for takeoff and, in normal circumstances, the last to land, as the tail was always the last to settle would jokingly remark that his pay should be higher because he spent so much extra time in the air! His lonely outpost called for all the dogged determination that he readily possessed. It was bitterly cold in his turret and the crew was always surprised that he could retain his cheerful disposition.

Despite the electrically heated flying suit, which had not failed him yet, the numbing cold penetrated to his very bones. The removal of the centre panes of Perspex from between the gun mountings did not help matters, but it did improve all around vision. If the suit failed, as they often did, dreaming about the warmth back home would see him through yet another bitterly cold night. The thought of smoking a bowl full of his favourite pipe tobacco as soon as they landed back at base also helped. He was well aware that most attacks on heavy bombers by fighters were made from the rear and this heightened his sense of responsibility. With excellent eyesight and good reactions, coupled with precise instructions to the pilot, he alone might be the saviour of the whole crew. So far, he and the rest of the crew had never even seen an enemy aircraft during operations, although like tonight they had often witnessed the results. This did not deter him from scanning the all-enveloping blackness for that elusive, dangerous shadow, which could mean certain death and destruction.

Suddenly there was an exchange of gunfire some distance ahead which almost certainly marked the fiery death of a Lancaster. Mesmerised for a fleeting moment, crew members watched horrified as it fell tumbling to earth. They could only hope that at least some of the crew had managed to escape.

Up front 'Pop' checked his watch. It was approaching 0151 and time to crawl below for the bombing run. With a fighter around perhaps he should wait just a little longer? No, he had his other job to do and needed to start the countdown to releasing the bombs. He and the rest of the crew hoped that one of the bombers' gunners had got a good shot at the marauder and sent him back to base damaged and licking his wounds, unable to rejoin the fray. But the German could still be out there, his injured pride serving only to heighten his resolve and sharpen his wits but like an injured animal he was more dangerous than before. The gunners redoubled their incessant searching. D'Ombrain knew only too well that they would be staring into the darkness looking for that enemy fighter. They were sure to fire first and ask questions later, calling out to him for the necessary evasive manoeuvre required the instant they saw something. But all was quiet again, save for the drone of the Merlins and the static hiss over the intercom.

Geoff's unsettled feelings began returning. He knew that each member of the crew had similar emotions and feelings, especially those whose homes and family were so far away. Yet Frank Reid was dreaming of the warmth back home. Geoff had listened to the German fighter controllers on the radio and jamming equipment had been set up to obliterate their signals. He could never understand a word that was spoken and would be able to do little about it, even if he could. In his mind's eye he could imagine the ground controllers positioning the fighters to intercept the bombers. There had been no recall signal earlier. They were unlikely to be recalled at this late stage. Geoff had passed on the briefed wind speed and direction for the flight and over the target that had remained unchanged. It would be some minutes yet before the Master Bomber gave them any instructions about the bombing of the target or any change to the briefed marker colours being used. Geoff set the dials to the allotted frequency, ready to receive any messages as they came in.[5] He had heard

---

5. The target area was apparently full of night fighters. The Master Bomber called for assistance but the anguished reply from the Deputy Master Bomber was, *I bloody can't. I've been hit*, whether by one of the numerous bursts of heavy flak in the target area at that time or enemy fighter activity. On the whole enemy fighter activity was considered intense in the target area. Presumed to be a relatively easy target, Trappes apparently lived up to its name and had become an operational aircraft trap. *Bomber Command: Reflections of War – Battles With the Nachtjagd, 30/31 March-September 1944, Vol.4* by Martin W. Bowman (Pen & Sword, 2013).

the final desperate shout from the other aircraft on his radio and knew that a fighter must still be operating out there somewhere.

Surely an extra pair of gunnery trained eyes would be helpful when searching the darkness for the enemy? Besides which, he might feel better watching the raid unfold and not just sit there. He unplugged his intercom and raised himself stiffly from his little seat. Climbing into the astrodome, he plugged into the intercom system, connected the oxygen and looked around. The outside temperature was around minus 20°. He soon felt cold and began shivering. How he wished that he had his Australian sheepskin vest with him. Perhaps a quick glance at 'Steve' pouring over his maps, a few shouted words above the incessant noise of the engines and then back to the relative warmth of the radio room would make him feel better. He had seen the markers starting to fall in the far distance and the first wave should be bombing now. He needed to be back at his wireless set in case any messages came through with instructions about the bomb run.

Although there was very little light in the radio room, it took only a few moments for Geoff's eyes to become accustomed to the dark. He could not see much below and the ground mist was obscuring most of the landscape anyway. What little light there was came from the crescent moon, but this was not enough to show any detail, especially from this height. Geoff had seen little of the French countryside from the air. Even when they had crossed France in daylight; once in the early morning when returning late from a raid and again when they had had an early evening takeoff he had seen only fleeting glimpses. His mind wandered back to the beautiful countryside of his island home that he loved so much. He thought of Muriel Gill, tucked up in bed at her home near Bournemouth. She was the only girl he really had deep feelings for. He hoped the Luftwaffe was not disturbing her slumber and he prayed for her protection. On his next leave he would ask her to consider joining him in the Isle of Wight when this war was over. He was planning to start a smallholding and raise chickens and pigs, perhaps even a few dairy cattle. First he and the crew would have to get this war won. Hopefully, she would have had the film of his crew that he had left with her developed. He had asked her to get seven copies printed, one for each of the crew.

No one will ever know what really happened in the bomber that night. The XV Squadron records simply show the aircraft as 'Missing. Nothing heard from 'C-Charlie' since takeoff.'

At 0150 hours 'C-Charlie' was suddenly filled with red-hot pieces of flying metal as cannon shells and tracer bullets ripped through the skin of the Lancaster! Geoff felt the aircraft give a shudder and lurch sickeningly downwards. No one had spotted Hauptmann Fritz Söthe's stealthy 'von unten und hinten' approach on his second totally unsuspecting victim of the night as he rose gently beneath the port wing of 'C-Charlie' and his aim had been perfect. Cloaked in the darkness, he throttled back to almost match the bomber's speed and then positioned his fighter perfectly in the blind spot. As he slowly came within range, still unseen, he opened the throttles and gave a gentle pull back on the control column. The Lancaster's port wing-root swung across his sights. Söthe squeezed the firing buttons with the deliberation of a sniper and carefully pushed the rudder bar with his left foot, raking the exposed underside of the wing and fuselage with one long burst from all his guns. All was utter confusion on the doomed Lancaster. Spent cartridge cases and ammunition belt clips showered onto the fuselage floor which was already awash with hydraulic fluid. When he had boarded the previous evening 'Sam' had removed the sack that was supposed to catch the spent shell cases. It made his entry in full flying kit into the confined turret that much easier. Unusually for him he had failed to re-attach the sack before takeoff. It was almost certain that 'Sam' was the first to be hit. The gunners had no chance to order their skipper to corkscrew to port and the German crew were past and away before any return fire could come close enough to hit them.

D'Ombrain wrestled with the control column and the rudder pedals but to no avail. Söthe's first salvo had done untold damage. The control system was wrecked and the aircraft refused to respond so the skipper gave the order, 'Bail Out!' But Lancasters were notoriously difficult to exit when flying fairly straight and level let alone when in a spin. As the bomber began its fiery final death spiral Söthe was sure that this would be confirmed as his 11th victory. He banked the 110 towards the east and calmly set course for Coulommiers where he could relax with a warming glass of Schnapps before retiring to bed.[6]

---

6. On 12 June Hauptmann Fritz Söthe was shot down and wounded in a Bf 110G-4 at Rozay-en-Brie. He and Unteroffizier Brönies were slightly injured and Unteroffizier Enke was severely injured. On 28 September Söthe was shot down and killed outright flying a Ju 88G-1 at Lambrecht-Neustadt; the victim of a RAF Mosquito nightfighter. Brönies and Unteroffizier Sabel were severely wounded and are believed to have subsequently died as a result of their injuries. In all, Söthe claimed eighteen victories, including one by day. He was awarded a posthumous Deutsches Kreuz in Gold on 1 January 1945.

'C-Charlie's wireless equipment and intercom system were smashed beyond use. Not that they were going to save them now. There was no outward panic; just the slow awakening that this was to be the final flight. None of the crew could escape now but Geoff was determined to try. However, even if he was able to reach the front escape hatch, he had no parachute. He had pulled back the curtain beside the navigator's table and tried to lean towards 'Steve' but his parachute harness snagged on some unseen projection as he squeezed along the narrow passageway and he was held firm. 'Steve' could just make out the contortions of his pal as Geoff wrestled with his parachute pack, who, in fear of pulling too hard and the pack opening in the confined space, tried to take off the harness but it was no good. Staring death in the face, all his earlier fears evaporated to be replaced by a calmness and acceptance of his situation. With certain knowledge that beyond the next few moments, there was to be no future and that all his dreams would never come true. Blinded by the acrid smoke and searing heat that rapidly filled the interior of the fuselage, visions of his past life began to appear in his mind. These were not like the shimmering black and white newsreel images at the local cinemas flashing and jumping across a smoky white backdrop before the main feature. These were in glorious Technicolor. At first there was only blackness. Then, much as the footlights would come up in a cinema, a pulsing sound and sensation accompanied a reddish glow. There was a rumbling that was felt as much as heard and yet feelings of floating inside some watery cavern were warm and comforting. These sensations ceased as a great pressure was exerted on his body and the reddish glow was replaced by white light. The screen curtains had been drawn. Gone were the watery sounds. Now everything was much clearer. Fluid in his lungs was replaced with air and was soon replaced by a new sensation. His body was being handled by something quite rough. With the air in his lungs and his throat cleared, his vocal chords were able to work and he let out a cry. Close by a soothing voice reached his ears.

There you are Elsie, a healthy bouncing boy, blond by the looks of it and everything's there alright. I'll just wrap him up and you can hold him.

Then another more familiar voice came through clearly.

Oh, he's so beautiful. George will be pleased, especially to know that he has a son and heir.

It was the voice of his mother, so long forgotten. Tears filled his eyes.

No other aircraft on the squadron reported seeing 'C-Charlie' go down but 7-year-old Bernard Postolle watching from a French farmhouse bedroom window heard the cannon and machine gun fire. Having grown up in the war and growing accustomed to the many aircraft at night he was quite adept at identifying an aircraft just by the sound of its engines and could even distinguish whether machine gun, cannon or Schräge Musik, which had a uniquely slower and deeper sound, was being fired. The youngster watched as the Lancaster came tumbling from the sky, passing just south of his father's farm and over the German fuel dump in the fields.

'C-Charlie' crashed and exploded on the gently rising ground less than one kilometre west of Lormaison, 20 kilometres south of Beauvais. Many of the villagers began to rush towards the crash site in a vain attempt to help any survivors but there was no hope. Aware that the villagers might try to spirit away any survivors, the German garrison billeted in the village and who used the magnificent Town Hall as their headquarters, herded all the French people back to the village and searched them to make sure they were not hiding a survivor amongst them. Bernard's father, who was the mayor of the village, remonstrated with the German Commanding Officer and was eventually allowed to return with him to the crash site, but only after the bodies of all seven crew were recovered and removed to the village cemetery to be buried. The local scrap metal dealer, Monsieur Letot, was impressed by the Germans to recover the wreckage of 'C-Charlie' and 'D-Dog' (ND926) on 622 Squadron at Mildenhall which had blown up in the air and had come down over a large area about 5 kilometres square with the bodies being scattered over a large area between Porcheux and La Houssoye. Letot collected the remains of both crews but mixed up the labels, resulting in Frank Reid being buried with those of 'D-Dog's crew. Later the bodies from both crews were exhumed and taken to Marissel French National Cemetery at Beauvais where they were subsequently laid to rest with the French Nationals. Their deaths robbed the world of many brave young men who had not lived to see the fruits of their labours. Families across the globe had sadly lost one of their relatives and friendships had been broken forever.

Chapter 9

Under the Influence of Mars

Francis 'Frank' Ridley Leatherdale was born on 3 November 1922 at Thornton Heath, Surrey. His father was an architect and structural engineer and became the Assistant Principal in the Architect's Department of the London County Council. Frank's mother qualified as a teacher at Homerton College, Cambridge and was a VAD (Voluntary Aid Detachment) nurse in WWI. Frank was the younger of their two sons. He and his brother Donald had the benefit of a good education augmented by a home environment which taught them to understand nature and the world around them. He had always possessed a greater than usual interest in war and martial matters, which started with playing with toy soldiers. His star sign is Scorpio and he was born under the influence of Mars. Perhaps there is something in astrology after all! In due course Frank was a first class rugby player and pistol champion. In describing his personal experiences he would ask you to remember that all his flying activity was as an air observer or navigator and as such he was the member of a crew who worked as a closely knit team.

Squadron Leader Frank Leatherdale DFC –
3 November 1922–23 March 2016

'In January 1944 our crew, which was skippered by Pilot Officer Donald S. McKechnie RCAF, who came from Ottawa and was a natural pilot left 1678 Conversion Unit for 115 Squadron at RAF Witchford, just outside Ely, where we joined 'C Flight. 115 Squadron had a proud history as a bomber squadron and was then equipped with Lancaster IIs. It was a big Squadron whose identification letters were 'KO' but it had used up the twenty-six letters of the alphabet with which to identify its individual aircraft. So 'C' Flight was given the identification 'A4A'. Our Flight Commander was Squadron Leader George Mackie DSO DFC – later to become Lord Mackie of Benshie – one of the few air observers to command a Flight, although he usually flew as a bomb aimer. He was every inch a Scot and his personality radiated confidence and the 'press on' spirit of the day.

'RAF Witchford was a typical wartime bomber airfield which began operating in June 1943. It had one long runway and two shorter ones forming a triangle. All the buildings were Nissen huts of various sizes, except for the two T4 hangars and the concrete Flying Control building. Personnel lived in sites dispersed in and around Witchford village. Station Sick Quarters and the three messes were on the north side of the village. After the war the Officers' Mess became the village school. Although as muddy as most wartime-built airfields, Witchford was a pleasant station; one advantage being that it was within easy walking distance of Ely.

'On 24 February 1944 we flew our first real operation, taking a 2,000lb MC bomb and a mixed load of incendiary bombs to the ball-bearing factory at Schweinfurt in southern Germany. It was a stormy night and the 7¼ hour flight was a severe test for me. But I was given no time to dwell on it for the very next night we attacked an aircraft factory at Augsburg – this time taking a 4,000lb 'cookie' as well as incendiary bombs. We had a full fuel and bomb load; not that this worried us at briefing. As the flight progressed the air became increasingly bumpy and there were thunder storms about. I became conscious of feeling the early effects of airsickness – a cold sweat and difficulty concentrating on the navigational calculations – this was nothing new to me I'm afraid. To add to my navigational difficulties the thunderstorms interfered with radio reception and Bernard – our WOp – could not get any radio bearings. The stars above were obscured by high cloud and the land below was invisible and so navigation was based only on 'dead reckoning' using the last wind velocity I had found as we crossed the enemy coast.

'I had adopted the technique of getting my crew to take off as soon as we were allowed so that we could climb as high as possible before having to set course over our base. My intention was to find a wind velocity as near as possible to our operational height as quickly as possible to see how it compared with the forecast w/v. On this night the next w/v I found was over the target which showed the wind to be a lot stronger than the forecast and I used this to set course for home. The thunder storms spread all across the continent. I still could not get any star sights to do astro-navigation and Bernard still could get no radio bearings with which I could fix our position.

'The night dragged on. Eventually I calculated that we should be near Witchford and 'Mac' called up on the R/T to get a turn to land but there was no answer. Bernard said that a lot of aircraft were calling the ground W/T stations asking for help, but they were unable to give any due to the storm. By now I should have been able to use the 'Gee' navigational aid which would have allowed me to fix our position by radar, but this too was impossible as the radar signals were being reflected from numerous thunder clouds and so cluttered the screen with spurious signals, I worked hard trying to strobe every combination in vain. We resorted to calling on R/T for help from 'Darkie'. With this system the aircraft called 'Darkie this is ... (your call sign)' and any airfield hearing your call would try to help you. The range of our R/T set was limited to nine miles to avoid interfering with other aircraft, so you knew you would be within no more than nine miles from an answering station. But this night there was no answer. We kept calling 'Darkie' as we flew on. We had let down from our cruising height of 20,000 ft sometime before, ready to land and to save fuel. All I could do was to give 'Mac' different courses to fly based on possible alterations to the wind. In the end I resorted to flying ten minutes north and then ten minutes west and then north again for ten minutes and so on. I was no longer trying to find Witchford; I was trying to find England!

'The dark, cold, night sky can be a very lonely and terrifying place when you are lost. Arthur, our flight engineer, had been worried about the amount of petrol we had left for some time and now he was checking and re-checking his readings of the contents of the various fuel tanks every few minutes. I kept trying to reason where we might be if the wind had changed in various directions and strengths; but still no radio call came from our base; or from 'Darkie'. Bernard said that the W/T was now blocked by so many aircraft sending out distress messages. 'Mac' had throttled back the engines to save fuel and was slowly loosing height. I was really quite scared – not for myself but I felt my responsibility to the rest of my crew. The fuel had nearly all gone and I warned the crew that I could not be sure whether we were over land or the sea and that they should connect their 'K' type seat dinghies in case they had to bail out over water.

'The red lights came on the fuel warning us that there was only a very little left. We knew that a Lancaster did not glide very well; when the engines cut it would go down like a brick and therefore we had to be very careful to bail out while 'Mac' still had control. I was just about to tell him that I thought we should abandon the plane when the rear gunner said he could see a searchlight shining on a layer of cloud below us to the south. We turned towards it and just at that moment a voice answered our 'Darkie' call. There was a lot of interference but I heard '…ford Bridge, standing by'. I carried a piece of rice paper (so that I could eat it if we came down in enemy territory) on which was written the code letters for certain main airfields, as well as the Very light colours of the day (which changed hourly), but there was no such airfield shown. However I thought it might be Stamford Bridge in Yorkshire as we had been flying north a long time. On the other hand we knew that the Germans had decoy airfields in northern France to which they attempted to attract lost RAF aircraft. However, we reasoned that it would be better to land there than to take a chance on bailing out into the unknown. As we broke cloud we could see an aerial beacon light, called a 'pundit' in RAF terminology, flashing in Morse code. However the letters were not on my flimsy and as we were now heading south we felt pretty sure that this was indeed a German decoy airfield. We carried a small incendiary bomb for use in destroying the plane if we landed in enemy territory. The bomb was to be stuck into a wing near a fuel tank and activated. We decided that the gunners would stay in their turrets after landing to fight off any enemy troops while Bernard got out through a hatch in the top of the cabin to set the bomb in the starboard wing. I carried a 9mm Mauser pistol and so I went to the rear door ready to help hold off any opposition.

'Mac' lined up on the approach to what appeared to be a normally lit RAF airfield. We touched down smoothly and as the tail dropped so the engines cut out having run out of fuel. I never want as close a call as that again. We trundled along the runway and in the light of runway lights and the searchlight reflected off the low cloud the gunners could see various aircraft parked around the airfield. They called out 'there's a Ju 88, there's a Messerschmitt, there's a Boston, there's a Halifax and so on. We were now quite sure this was a German airfield. I was fumbling in the darkness to open the rear door when it was opened from the outside and a very English voice said: 'Welcome 115 Squadron'. He was obviously a good old RAF 'erk' out there; and who else would know that our unusual code letters of 'A4A' were 115 Squadron. I put my pistol behind me while

I found out more. It transpired we had landed at Hartford Bridge (now known as Blackbushe) near Guildford, in Surrey. It was the only airfield open in the south of England that night and its searchlight shining straight up on the clouds was attracting aircraft from all over Bomber Command. The presence of the German aircraft which our gunners had seen was due to the fact that this was the main base for the RAF's 2nd Tactical Air Force and pilots were trained there to fly German aircraft ready for the invasion of Europe, so that they could fly captured planes back to England.

'There were many tens of bombers arriving and the airfield's facilities were swamped. We were de-briefed and a message sent to inform Witchford that we had landed safely. After a meal we slumped into chairs, or on the floor, in the Mess; as all the beds had long been occupied. The next morning daylight revealed the incredible activity of the night before. There were Halifaxes from Yorkshire, Lancasters from all over Bomber Command and some night fighters parked all around the airfield and on the grass. It had been so hectic that Flying Control had lost count of the number of planes that had landed! We were told it would be at least two days before the limited ground crew at Hartford Bridge would be able to service our plane and a further delay would be caused while they got in supplies of petrol for us.

'As I lived only a bus ride away at Leatherhead I took 'Mac' home with me and gave my parents quite a shock seeing us arrive on the door step dressed in our flying kit. We stayed the night with them before returning to Hartford Bridge the next day. Our Lanc' had been refuelled and we flew back to Witchford. I was still shaken from my experience and felt that I should ask to be taken off navigational duties, as I blamed myself for risking everyone's lives. However the Station Navigation Officer, Squadron Leader Klufas RCAF[1] said I had done all I could, that hundreds

---

1. William James 'Rip' Klufas, one of the highest ranking Ukrainian-Canadians to serve with the RCAF, was born in 1915 at Radway. His parents were John Klufas and Mary Klufas (Lazowski), immigrants who arrived in Canada in 1910. Of six children born into the family, three sons became teachers and served in the Canadian armed forces during the Second World War. 'Rip' obtained his permanent teaching certificate in 1936 and was teaching when he signed up with the RCAF in 1941. He joined RAF Bomber Command overseas as a navigator and rose to the rank of Squadron Leader and later Base Leader. William's brothers Harry and Peter served in the Canadian Army, the former with the Canadian Military Headquarters staff in London.

of aircraft had been caught out by the storm and the winds had differed completely from those which had been forecast. Most crews had been blown a long way south and a bit east of their estimated positions. The members of my crew said they did not want to fly with anyone else as their navigator. Their confidence in me was heartening, but I don't think I shared it. I was very worried.

'The following night we were again on the order of battle. This proved to be an even longer trip, to take a 4,000lb 'cookie' and a mixed load of incendiary bombs to Stuttgart. Things went according to plan and when we landed after seven hours and 35 minutes I was feeling happier and more confident in my ability as a navigator. After this operation we were given seven days leave. At this point John Wagner, our rear gunner, decided that he could not endure flying anymore. He was now in his mid-thirties and was considerably older than most aircrew. This was not a case of LMF – Lack of Moral Fibre – which was the term used to describe when someone's nerves failed. John was just unable to stand the extreme cold for so long on each flight. He was grounded and transferred to administrative duties, for which he was eminently suitable having been a bank manager.

'On my way back from leave the train was late getting into London and I missed the last train from Liverpool Street station and I had to spend an uncomfortable night in a carriage waiting for the train to leave at 4.30 am and all the time keeping half an ear open for the sound of any air raid alert.

'On reaching Witchford I found we were on the battle order. As I pasted my charts together before the main briefing I had few worries until I realised that the chart labelled München was really referring to Munich. How green I was! I knew a Lanc' II could not reach Munich and return. We spent a very worried afternoon awaiting the main briefing. Some old hands said that if Munich really was the target they would go to Switzerland after bombing and get themselves interned there. After our Augsburg experience I was very worried, perhaps even 'afraid' might be an appropriate description. Luckily on this occasion the planners realised their mistake and the Lanc' IIs were taken off the order of battle. Later on we did hear rumours that at one of the Canadian Lanc' II stations there had been a near mutiny before the cancellation came through.

'The next night our target was the big one – Berlin. After our recent experience this might have been worrying but for the leadership shown by our Flight Commander – Squadron Leader George Mackie – who put himself on the battle order, making it his eighth trip to the 'Big City'. This was typical of the man and most reassuring to the rest of us.

'For our next four operations we were given different rear gunners until a regular replacement for John Wagner could be found. This night our rear gunner was 'Chuck' Koss, an American 'Top' Sergeant who had joined the RCAF before the USA entered the war, but after Pearl Harbor he transferred to the USAAF and so enjoyed a greatly increased rate of pay. However, he wanted to remain with his crew and the RAF and so here he was. He wore USAAF khaki uniform and he habitually had the peak of his 'doughboy' hat turned up looking like a cartoon character from the Bronx.

'We took a 1,000lb HE bomb and 5,040lbs of incendiaries which gave us a weight allowance to take a full load of petrol. The route was straight out across the North Sea, across Denmark and over the Baltic before turning almost due south for Berlin; then we were to turn west and across Germany passing north of the large defended area of the Ruhr valley. At 20,000 ft the Met forecast a North-westerly wind increasing in strength as we went eastward to a speed of 45 knots. Following my crew's usual plan we took off as soon as we were allowed and climbed for height while circling base. As a result I found the wind was already stronger than forecast by the time our 'Gee' was jammed by the Germans as we got about halfway across the North Sea. As we crossed the Danish coast Ken, our bomb aimer down in the nose, got a pin point which showed we were well south of track. He was positive that he definitely recognised where we were from his map. This gave me a wind from the North West but at 65 knots. I used this for navigation. We did not see the ground again.

'At the time I calculated we should be over Berlin there was no sign of enemy searchlights nor flak, or of our Pathfinder's Target Indicator (TI) markers. Then suddenly all hell broke loose behind us as the first PFF flares started to cascade and the defences opened up. Doubtless they had been silent hoping we would fly past them to go somewhere else. The ground was hidden by a solid layer of cloud so the Path Finders were dropping sky markers. These were special parachute flares which burnt with a given colour, either red or green and sometimes they emitted stars of a different colour which fell from the main flare in order to distinguish our markers from German decoy 'sky markers' fired from their anti-aircraft guns. The PFF aimed their sky markers using their H2S radar, but for the main force to hit the target on the ground the bombs had to be dropped on a given heading. In our case it meant we had to turn round and fly back across Berlin before we could turn again onto the correct heading for bombing. As we flew back it took us twenty minutes to fly northwards across Berlin. Unbelievably, our ground speed was only 55 mph for the wind was now blowing at 120 mph or more. It was quite frightening as we risked colliding with friendly aircraft coming towards us in later waves and we must have been a sitting duck for the flak; but our luck held and we safely got into position from which to bomb. I used the new wind I had found to re-calculate our course for home. Soon after leaving the target our main compass failed, which meant that 'Mac' had to steer by the P4 compass. This was difficult for him to see and was not so accurate. We got back to Witchford without further trouble, having been airborne for seven hours and 25 minutes. We subsequently learned that Bomber Command had lost over 100 heavy bombers. Some had been blown off course and had straggled across the Ruhr where the defences had a field day. Others, like us after Augsburg, ran out of fuel over England and crashed. George Mackie had come down and flew below the clouds, map reading his way along the Dutch coast.

'Berlin was also our last operation in Lancaster IIs, as these were now replaced by Lancaster Is and IIIs; both of which had Merlin engines, the difference between them was that the Lancaster I had Rolls Royce built Merlin XX engines, while the Lanc' III had American Packard built Merlin 28 engines with Stromberg carburettors. The ground crews liked to have Lanc' IIIs as each one came complete with a full American tool kit. On 30 March we had been allocated a brand new Lancaster III (ND758), which bore the letters A4A. The crew nicknamed it The Bad Penny because bad pennies always come back and the ground crew painted this name on the port side of the nose. She went on to complete over 100 operations and was finally scrapped after the war. Whilst we regarded it as 'our Lanc' other crews flew it when we were not on the battle order. On our first familiarisation flight 'Mac' quickly discovered the big difference between the takeoff power of Merlin engines compared with Hercules and we failed to gain height in the way we were accustomed and flew low over the Ely to Cambridge road to the surprise of people on the ground and to those in the Lanc' who could see out! Two of our familiarisation flights when 'Mac' was getting accustomed to our Lanc' III were height tests with a full bomb load. On the first occasion we reached 26,000 ft and the next day we got up to 27,000 ft. This was the highest I ever flew in a Lancaster.

'On 9 April we attacked the marshalling yards at Villeneuve-Sainte-Georges near Paris and on 11 April we attacked those at Laon. Although we did not know it at the time these attacks were preparing the disruption of the German rail transport system ready to stop them sending reinforcements to the Channel coast when it was invaded on D-Day. Our bomb load to Laon was ten 1,000lb and four 500lb HE bombs. This made us very heavy for takeoff. In a Lanc' II there would have been no problem, but it was different in a Lanc' III and to make matters worse we were using the short runway for takeoff. With full power we just managed to stagger over the spires of Ely cathedral.

'As with all the targets in France we had to take care not to drop bombs outside the exact target, but at Laon we were told to be particularly careful and not to bomb if the bomb aimer was not sure of a perfect hit. The reason was that the marshalling yard was right beside the town. The railway was down on the plain while the fine old town of Laon stands high above it on a bluff. When I visited Laon after the war I could see very clearly why we had to be so careful on this target. Even when we had got back to base we were not out of danger. With so many aircraft flying around in the dark, either with no navigation lights at all or with very dim ones, there was always the risk of a collision. At Witchford our circuit actually overlapped that of RAF Mepal, where 75 New Zealand Squadron was based. It was not unknown for a pilot to land at the wrong airfield. As we got near base on our return the pilot would call on R/T asking for his turn to land. Usually this meant stacking – you were given a height at which to circle base while aircraft who had returned ahead of you landed. This we called 'orbiting the beacon' (the beacon, more properly called a 'pundit', being a red light near the control tower which flashed the Morse code identification letters of the airfield.)

'But the perils of night flying in crowded skies were not our only worry. Around this time the Germans became increasingly active in sending their night fighters in with the returning bomber stream to attack our airfields or any aircraft they could find. A Lancaster with its undercarriage down and 15° of flap down was an easy target. These night fighters were often undetected by British radar as they merged with the bomber stream. Our gunners had to remain very alert, ready for instant action, until the engines were switched off in dispersal.

'One night, we were approaching Witchford, and could hear other aircraft being given their heights and turns to land when our gunners saw a combat taking place. The thin red lines of RAF tracer were quite easily distinguished from the German cannons which used white tracer. The Lanc' involved managed to fight off the attack this time, but our airfield was still illuminated. War-time lighting was very subdued but from the correct angle you could see the Drem circle (lights which indicated the circuit) and the runway lights. A very distinctive Australian voice came over the R/T 'Put those lights out there is a bandit in the circuit'. Flying Control answered saying there was no intruder. By this time we had seen a second, inconclusive, combat. The Aussie voice called out 'If you don't put your bloody lights out I'll shoot them out'. The airfield lights were switched off. The orbiting aircraft continued to circuit awaiting the intruder's departure. Sometimes when intruders were near base we would be sent on a short cross-country flight until the danger had gone. That night we all landed safely.

'It was not always so. On 19 April two Me 410 night-fighters got into Witchford's circuit undetected and in the space of twenty minutes shot down two Lancasters: Flight Lieutenant Charlie Eddy RNZAF and crew on 'R-Robert' (LL667) and Pilot Officer John 'Jock' Birnie and crew on 'J-Jig'(LL867) – all the crew members were killed. Both Lancasters were caught in the 'funnel' that led from the Drem circle lights to the runway itself as they turned onto their final approach to the runway and they fell half a mile apart into very soft soil in Coveney Fen.[2]

'We were now sent Sergeant Joe Hayes, from Wrexham to replace John Wagner as our rear gunner. Not only was Joe much younger, but he was also much smaller and so fitted into the confined space of the

---

2. Charlie Eddy (29) was an Australian, the son of Matthew Ernest and Isabella Eddy, Hamilton, Victoria. His Lancaster was shot down by Oberleutnant Klaus Bieber of 5/KG51. Eddy is buried at Cambridge City Cemetery. Twenty-year-old Jock Birnie's Lancaster was downed by Major Dietrich Puttfarken of 5/KG51 who was himself killed four nights' later on 23 April on another 'Intruder' sortie over eastern England.

rear turret rather better than had John. Our crew was complete again. On 24 April we took a 4,000lb MC 'cookie' and 4,320lbs of incendiary bombs to Karlsruhe, our first operation with Joe in the rear turret. This flight lasted 6¼ hours and once again we had no enemy action directed against us.

'On 26 April we attacked Essen which, in spite of it being in the well defended area of the Ruhr we managed without undue incident and were airborne for only 4¼ hours. The next night we went to Friedrichshafen on the shore of Lake Constance. This was a long flight lasting 7¾ hours. The target was very close to the Swiss border and we were warned that the Swiss would fire on any aircraft flying over their territory, if only to prove to the Germans that they upheld their neutrality. Therefore we had to make a much tighter turn than usual after we had released our bombs. Our gunners saw Swiss ack-ack guns firing and we later heard that they had shot down one plane.

'We were again flying with different rear gunners as Joe had suffered frostbite to his cheeks when the metal clips on his oxygen mask touched his skin on the raid on Karlsruhe. These clips should have had chamois leather on the side nearest your face, but Joe did not notice that his was faulty. Thus when we went to attack the railway yards at Chambry on 1 May we once again had the American Top Sergeant 'Chuck' Koss with us. Although this raid took only three hours and twenty minutes it was far from a 'piece of cake' – operations seldom were. This time we had some predicted flak fired at us and we had several holes for the ground crew to patch. You could hear pieces of shrapnel rattling on the Lanc's metal skin. It was a distinctive sound which I will never forget.

'About this time Flight Lieutenants Halley and Seddon from Newfoundland and Australia respectively completed their operational tours of thirty raids each on the same night. In fact they were the first crews on 115 Squadron to complete a tour of ops for over three months. The Air Officer Commanding 3 Group was waiting at Witchford to award each an immediate DFC. There was no doubt that the Squadron morale rose after this event and never looked back.

'Our Lancaster (ND758) was fitted with H2S radar. This was both a navigational and bombing aid. The equipment provided a sort of picture of the ground below and once the operator had become experienced in interpreting the signals it was very useful. Water surfaces such as rivers or coastlines showed up the most clearly as dark areas with no radar return signals. The most prominent returns came from buildings in a flat area, but sometimes the returns were masked by being in the radar shadow of hills. This was the radar with which the heavy bombers in the Path Finder Force were equipped, but now there were sufficient sets coming off the production line to equip some of the main force aircraft. We were indeed fortunate to be chosen to have one as there were only three such aircraft on 115 Squadron at that time.

'On 8 May we took a 4,000lb MC 'cookie' and sixteen 500lb HE bombs to attack the airfield at Nantes at the mouth of the Loire. This airfield was being used by Focke-Wulf FW 200 Condor long range bombers to attack shipping in the Atlantic and to work with U-boats and so this was a high priority target at the request of the Admiralty. We landed after 5½ hours in the air. Then on 10 May we took fourteen 1,000lb HE bombs to attack the coastal guns at Cap Gris Nez, although the moon was one third full we had no trouble and landed after just two hours and five minutes. We had taken off at 0255 hours and were back in bed by 0630. I got up at midday and found we were on the battle order again that night and we took off at 2205 hours to take fourteen 1,000lb HE bombs to attack the marshalling yards at Courtrai. This proved to be another short trip lasting two hours and 25 minutes. We had flown two raids in one day!

'On 19 May we attacked the railway at Le Mans with a mixed load of HE bombs as part of the softening up process before D-Day, during which we attacked railway junctions and other targets on the German lines of communication. There were a number of night fighters operating around us resulting in several combats, but we were unscathed. This op lasted 4½ hours and I used astro to aid my navigation on the flight home. Two nights' later we went to Duisburg at the western end of the Ruhr with a 2,000lb HE bomb and a mixed load of incendiary bombs and were airborne for 3¾ hours. We had no real trouble and it began to look as if the tactics we had evolved were paying off. The planners at HQ Bomber Command routed the bomber stream to avoid known areas where anti-aircraft guns were concentrated and also so that we flew through the area defended by ground controlled night fighters as quickly as possible.

'The night after Duisburg we were again in the Ruhr, this time attacking Dortmund. I was one of the navigators detailed to wireless back to HQBC the winds I was finding, as part of the zephyr wind procedure for the other bombers. Now that we had H2S I had few qualms about being given this responsibility. We took a similar bomb load as we had taken to Duisburg, but this flight lasted four hours. On 28 May we took ten 1,000lb and five 500lb HE bombs to Aachen. This was not only a railway target, but also the Germans had their main anti-aircraft gunnery school there. So while we expected a hot reception, we had the satisfaction of knowing that we were getting back at the flak gunners. This proved to be an uneventful raid lasting 3½ hours. At the end of May I went to RAF Waterbeach for a night vision test; and then we were given seven days leave from which we returned on 6 June – D-Day. We found ourselves immediately on the battle order, which, of course, was where we wanted to be on such an exciting day. That night we attacked the railway yards at Lisieux. Most unusually we were ordered to fly at 3,000 ft so that our bombing would be even more accurate as it was so important that this rail point should be denied to the Germans. At this height the air was very bumpy and I was airsick, however it only lasted for just over three hours and I would have endured anything knowing we were in active support of the British Army.

'We had another target in support of the invasion on 8 June when we attacked the marshalling yard at Chevreuse, near Paris. This was quite a 'hairy' trip and our port inner engine was hit by flak but we got back to base safely. 115 Squadron lost six Lancasters attacking this target, with three of them from 'C' Flight. At the time Eric reported seeing lines of tracers coming up from below and we thought this must be from light AA guns, as we were only at 5,000 ft. However French sources have since told us that most of our aircraft were shot down by night-fighters. Apart from our losses other Squadrons lost 28 aircraft attacking communication targets in the same area that night. In addition four Halifaxes from 138 Squadron at RAF Tempsford were lost flying somewhere below us trying to drop supplies to the Resistances fighters. Most of our aircraft crashed in a fairly small area to the west of Paris – it seems we had stirred up a hornet's nest. We now wonder if what Eric thought was light flak was in fact tracer cannon fire from night-fighters equipped with upward firing guns – a system the Germans called 'Schräge Musik', about which we did not know until after the war. Usually the Germans would not have used tracer shells with 'Jazz Music', for fear of revealing their position, but relied upon getting in very close to the bomber before opening fire from below and aiming at the exhaust flames from the inner engines and so trying to hit the main fuel tanks in the wings.

'On 10 June we again attacked a railway marshalling yard, this time it was at Dreux. It was my custom to memorise the next course to give the pilot after we had bombed and then to come out from behind my blackout curtain and stand behind the pilot and flight engineer to help to look for fighters and other aircraft. On this night I was doing this as the attack started to develop. We heard the Master Bomber (whose call sign that night was 'Little John') broadcast to the main force (us) 'Main Force from Little John bomb.' He then went off the air and as he did so I saw an aircraft above us out on our starboard side having a combat with a night fighter. At that moment I could not see the aircraft, but I could see the thin lines of red tracers from an aircraft's four rear machine guns and the much fewer big balls of white tracer from the fighter's cannon. Then the cannon fire began to burst on what I could now see was a Lancaster, there was several explosions and the Lancaster began to burn from wing tip to wing tip and from nose to tail and it climbed up and dropped its port wing as it came over above and slightly in front of us before spiralling down. As soon as 'Little John' cut out we heard the Deputy Master Bomber calling 'Little John from 'Little John 2', 'Little John' from 'Little John 2'. There was no reply and quickly came the message 'Little John 2' to Main Force, taking over …' and then he too cut out. In almost the same part of the sky I saw another combat but this time the balls of cannon tracer were coming up almost vertically and soon there were sparkles at the end of the tracers and I saw a Lancaster being attacked with cannon fire from a single-engined Focke-Wulf 190. The FW 190 was very close to the Lancaster and raking it from nose to tail and back again as it climbed very steeply underneath it. I had to admire the skilful flying performed by the German pilot in this 'Wild Boar' fighter. 'Wild Boar's were not being controlled by GCI but were attacking such targets as they could see in the dark. With both the Master Bomber and the Deputy Master Bomber being shot down almost together we carried on to bomb as we thought best.

'The next night we were sent to attack the airfield at Nantes once again. We carried eighteen 500lb HE bombs. As we approached the target the Master Bomber called the Main Force down from our height of 15,000 ft to below a solid layer of cloud at about 4,000 ft. The airfield was lit up as if in daylight by the light of the many searchlights being reflected off the cloud base. All hell was raging with the sky criss-crossed by tracers from German AA cannon and machine guns and tracers from our aircraft firing back. From this height you could not miss and Ken aimed his bombs across the line of hangars housing the Focke-Wulf Condors. We were being shot at by some accurate light flak and their shells started to burst on the port tailplane and Eric had some perspex shot out of his turret and he shouted 'Dive!' to 'Mac' to get away from the string of shells. But 'Mac' did not immediately understand what had been said. Normally in our crew we insisted on strict intercom discipline and when you spoke you would always say who you were and to whom you were speaking – like 'navigator to pilot…' On this occasion there was no time for such niceties. 'Mac' started to push the control column forward but Arthur realised 'Mac' had not understood what Eric had said and he reached over to push the column as well, resulting in the control column being pushed forward even more vigorously. All my navigation instruments went flying off my table. The flak scored more hits on our starboard wing and near Ken in the nose before we evaded it. After we left the airfield we turned west out over the mouth of the Loire and Ken saw a large merchant ship which he could have bombed had he had a bomb left. We returned to England over the south Devon coast which looked very peaceful in the early light of dawn with wisps of smoke coming from some chimneys of houses where early risers were obviously awake. We were still at a fairly low level and I wondered what the people below were thinking, no doubt woken by us roaring overhead. We landed after 5¼ hours.

'On 15 June we delivered eleven 1,000lb and four 500lb HE bombs to the docks at Le Havre which was another short trip which hardly took us over enemy territory. On 18 June we had one of very few abortive trips when cloud over Montdidier railway yards prevented Ken from seeing the aiming point and we had to bring our bombs back. This was never a popular thing to do for both the aircrew and the ground crew. The pilot had to contend with landing an aircraft which was much heavier than he would normally experience; and the ground crew had the task of taking the bombs off. This was arduous work, winching each one down and putting it back on a bomb trolley but it was also quite dangerous. Most bombs had delayed action fuses fitted to them which worked by a glass phial of acid being broken. The acid then ate its way through a celluloid disc which was keeping the striker from being pushed by a spring onto the detonator. The length of time the delayed action took to work depended upon the thickness of the celluloid discs. These delayed action fuses were not intended to operate until the bomb had fallen a certain distance, during which the wind would unwind a small propeller allowing the arming device to operate. However the glass phials were sometimes broken by other means – such as a severe jolt. Hence the armourers needed to use extreme care.

'The 23rd June saw us making our first attack on a V1 flying bomb site when we bombed L'Hey. While we plastered the area with eleven 1,000lb and four 500lb HE bombs it was a very difficult target to hit at night. L'Hey was one of the places where V1s were assembled and serviced, so it presented a larger target than the actual V1 launching ramps but it was still quite small. We were guided by PFF Mosquitoes using 'Oboe' to mark the aiming point. In this situation 'Oboe' was a very accurate system which used ground stations in England to transmit signals to guide the Mosquito pilot and to advise him when to drop his TI. This proved to be our shortest operation ever, taking only one hour and 55 minutes from take off to landing.

'On 27 June we were operating in support of the Army again. We bombed the railway at Biennials, dropping eighteen 500lb HE bombs across the railway lines. Many bombs had delayed fuses so that they would go off at varying times after landing and so disrupt the German's attempt to repair the damage of those bombs which had exploded instantly. On 30 June we had been gathered in the briefing room, as usual and found ourselves on the battle order. The navigators were issued with the blank charts for the coming night and had to paste them together to cover the whole route to and from the target. We did this in the morning so that the paste had time to dry before we had to use the charts. At the same time the gunners were helping the armourers to belt together the .303 ammunition, putting tracer rounds amongst the ball bullets to suit the gunner's requirement. Some of the pilots would be practising instrument flying and blind beam approaches in the Link trainer. From the area covered by the charts which had been issued and knowing what the bomb load and fuel load were some of the older hands could make a shrewd guess at what the target would be. On this day it looked as if we were changing from bombing railways in northern France in direct support of the land battle and were returning to attacking the German war effort, as our target was obviously in southern Germany and looked like being a large town as we had a mixed load of HE and incendiary bombs.

'We now went to bed to sleep until the early evening when we would get up, have a pre-flying meal and go to the main briefing prior to getting into our aircraft. However our sleep was disturbed at about 2.00pm by the Tannoy announcing 'All aircrew report to the Briefing Room for a lecture.' We knew this could not be right, as we would not have been disturbed prior to night operations for a lecture. We hurriedly dressed and cycled hell for leather up to the airfield. There we were met with a surprising scene. The usual pre-operation busy routine was even busier with bomb trolleys going in all directions, some loaded but others empty. In the Briefing Room our CO – Wing Commander Guy Devas – who had recently taken over from Wing Commander Bob Annan – said: 'Navigators stay here, all other aircrew go to your aircraft; take your flying clothing with you and help the armourers change the bomb load.' All the incendiary bombs were to come off and be replaced with HE bombs.

'We navigators were now told the details of the operation. Word had been received that the Germans were at last reacting to the invasion and were moving a Panzer division up from southern France. The French Resistance were doing what they could to delay the movement of the Panzers on the railway, but this only slowed their movement a little. Their imminent arrival with their latest mark of Tiger tanks could push our lightly armoured troops back to the sea. On the Panzers' route lay the town of Villers-Bocage. All roads went through it and the Panzers could move through the Bocage countryside only very slowly as the small fields were each surrounded by banks along the top of which grew thick trees and hedgerows. It was not good tank country. We were to bomb Villers-Bocage to make it an impassable barrier to the Panzers. 'Monty' had sent an urgent appeal to Air Marshal Harris for all possible help and Harris lost no time in answering the call.

'Wing Commander Devas said we were to take off as soon as our planes were ready – no waiting for a given time, no 'H' hour at the target – we were to fry direct to Villers-Bocage in a straight line. There would be American Thunderbolts providing fighter cover below us; and RAF Spitfires would provide high level cover. We were to fly at 14,000 ft as at this height the mark XIV bombsight was at its most accurate. This was our first daylight operation. The CO suggested we get ourselves into vics of three as we took off so that we would have some support from the other planes' machine guns. The first Lanc' off would lead, and the next two, would formate on either side. Then the fourth Lanc' off would lead the next vic and so forth. Our Lanc' was loaded with eleven 1,000lb and one 500lb HE bombs. There had not been enough time to remove any petrol, so we had all the weight of full tanks which we would have needed if we had flown to southern Germany. Now it was just dead weight. We took off at 1755 on a clear sunny afternoon, led a vic and were soon climbing as we flew towards the Channel. Eric and Joe were reporting all manner of aircraft around us – other Lancasters of course – but also Halifaxes, Wellingtons and various fighters.

'As we crossed the Channel we could see ships of all shapes and sizes heading north or south with their wakes streaming out behind them, white lines in a stormy white flecked grey sea. It was indeed a truly wonderful sight. As we crossed the Normandy coast we saw a formation of Halifaxes ahead and above us. 'Mac' decided to lead our vic in a little below them because the Halifax had four guns in its mid-upper turret, compared with only two in a Lancaster. We had to keep a bit behind them to avoid being hit by their bombs when we reached the target.

'As Villers Bocage came into sight I left my 'office' and stood in the cabin. There were some black puffs of flak smoke over it. There was no sign of fighters – ours or theirs. Then one of the Halifaxes was hit. There was no time to dwell on the fate of its crew for we were now on our bombing run. I wondered what all the red smoke was that was now obscuring the town and then realised it was brick dust from the exploding bombs dropped by the first wave of Halifaxes. I had never seen bombs bursting in daylight before. As we turned to the west and round to head for base Ken saw a battery of guns drawn up behind a hedge. They were pointing northwards so they must have been German. Not for the first time Ken wished he had had a 'hang up' so that he could have had a go at bombing them as they presented a good target. Sometimes its release gear might be frozen, or perhaps there had been a temporary electrical fault. Or even friction might have prevented the release gear working. Whatever the cause you could often get a hang up to drop at a second attempt and it was this type which Ken could have utilised. This raid had taken three hours and ten minutes and was probably the most sensational operation we ever carried out. After we had landed Eric told me that he saw a shell burst between the port engines of the Halifax and this shot off the propellers which then hit its wing.

'All our flights in June had been operational sorties and we had now been credited with 27 operations. 'Mac' had done one more op than the rest of us as it was the custom for new pilots on a squadron to fly their first operation as co-pilot to another more-experienced crew. I applied to our Wingco to go to the Path Finder Force when our tour was complete. At first 'Mac' said he had done enough and would go back to Canada. Our flight engineer, Arthur France, was by now married and he too did not wish to continue flying operationally. My application for service with the PFF was accepted.

'On the night of 7/8 July we took ten 1,000lb and two 500lb HE bombs to the marshalling yard at Vaires near Paris; and on 10 July we had a daylight operation when we took off at 4.00 am to take eleven 1,000lb and four 500lb HE bombs to Nucourt. This was another flying bomb assembly plant in Northern France. The target was covered by cloud but we had been ordered to bomb using our 'Gee' equipment if this was the case. At that distance from our signal stations and now that the German jamming of 'Gee' had been curtailed, we had good signals; but all the same it was asking a lot to hit so small a target through clouds using 'Gee'.

'So our first tour of operations had ended and we were all unharmed. I experienced mixed feelings of wild excitement and of anti-climax. I had flown thirty operations (including our 'Nickel'), carrying 284,160lbs of bombs and covering 23,710 miles in 135 hours. I still trembled when I thought of our flight from Augsburg but I had gained a lot of experience and knowledge along the way and now felt I could handle most situations. By now Squadron Leader R. H. Smith had replaced Klufas as the Squadron Navigation Officer and he wrote in my Flying Log Book that my proficiency as a navigator was 'above average' and that I was 'keen and very efficient. 'Mac' was awarded the DFC.

'A new chapter was about to begin. 'Mac' and the rest of our crew except Arthur now said they would like to come with me to Path Finders. At the beginning of August 1944 we reported to 7 Squadron at RAF Oakington, near Cambridge, which was a Path Finder Force squadron equipped with Lancaster IIIs and we were immediately sent on detachment to the Path Finder Force Navigational Training Unit at RAF Warboys, near Huntingdon. Before we left Oakington we were given a new flight engineer to replace Arthur France. Flight Sergeant Stephanos Haralambides was a Greek Cypriot marine engineer who had been in England studying for a marine engineer's exam when the war came and he joined the RAF. 'Steve', as he was known, was a fine engineer, which he quickly proved to me by tuning the twin carburettors of my Singer Le Mans car so that its performance improved beyond all recognition. We sometimes had a little difficulty understanding his guttural voice – I'll always remember his 'Changing tonks now skipper' coming through the intercom.[3]

---

3. Squadron Leader Francis 'Frank' Ridley Leatherdale DFC flew 59 operations with Bomber Command, 29 of these on Pathfinders. He passed away on 23 March 2016.

Chapter 10

Kaiser's Crew

In front of us out of the nose there was a mass of black spots all over the sky; flak! I felt that I could get out and walk on it at 19,000 ft, a peculiar sensation and I never forgot it.

Flight Sergeant Bill Wooldridge, flight engineer on Flight Lieutenant Jack Kaiser's crew on 90 Squadron in 3 Group RAF at Tuddenham in Suffolk, which crews called 'Muddenham' because of the rain.

In rural Saskatchewan in 1940, a sixteen year-old farm boy was out shooting gophers when an aircraft flew overhead. That was the day that John William Kaiser decided to be an aviator. 'Jack', as he was known to family and friends, was born in Harris, Saskatchewan on 28 June 1924, brother to three siblings. Raised as a farm boy he also learned mechanics in his dad's car dealership. He curled and played hockey in the local rink. After seeing that aircraft, he 'borrowed' his older brother's ID and Jack eventually enlisted in Saskatoon in 1942. On 4 August he set out on the train from Saskatoon to #3 Manning Depot Edmonton. Unfortunately a couple of men from Saskatoon 'tried him on' in the wash room of the train. They had made a mistake. Jack had worked in a gravel pit most of the summer and as such was hard as nails. Also he was worrying because when he left home, it was the first time he had seen his Mother cry. One of the men called Jack a 'hick farmer', which he was but didn't think so. Urged on by his mate he kept on about 'hicks' so Jack told them, as he was drying his face after he had shaved, 'to pack it up'. His answer was 'who is going to stop me?' So Jack hit him and he fell and banged his head on the iron support of the seat. The other man rushed Jack so he stuffed his face in the sink and turned on the hot water. They never bothered him again. One made it to officer status as a gunner and survived. Jack met him after the war on the street in Saskatoon and they went and drank some beer together. Jack reminded him of that train trip in 1942. He never knew what happened to the 'stupid' one.

Jack completed his initial flying training at Regina and elementary flying training school at Virden. He went solo in a Tiger Moth on 21 February 1943 nearly a month after his arrival. The propeller came off his aircraft on the downwind leg. He shut down the engine – 'it was sure quiet in there' and watched the prop land in a snow bank, turned onto base leg, followed the Tigers in front and landed. Jack wanted Fighter Command Spitfires in England. He knew he must do as well as he could so he set himself to work and completed his course, standing first in Ground School (average 93%) and both flight tests (pilot and navigator) were marked 'Above Average'. He fully expected to be posted overseas, because he had written 'fighters – ASAP' but he was posted to General Reconnaissance School in Summerside, the entry school for Maritime Command. He was, to say the least, 'pissed off' and said so. Eventually, he was sent to Halifax Holding Unit and posted to England, where he boarded a train to the Manning Pool at Bournemouth. Christmas came and went and in late February he went on a bus with about thirty others to Coventry. They toured the war-torn city and returned to Bournemouth with a totally different outlook. Jack had made up his mind. 'Screw the fighter pilot vision', give me a chance to help win this war, not as a defender but as an attacker.' And he wanted the most powerful weapon they had – the mighty Lancaster. He got his wish.

On about 21 March 1944 he was posted to No.12 OTU at Edgehill where he felt extremely lucky to get Flying Officers' Ross Meggason and Howard Keon and Sergeants' 'Jimmy' Ginn and 'Pete' Campbell, who were bomb aimer, navigator, wireless operator and mid-upper gunner respectively. 'Howie' Keon was born 14 October 1920 in Haileybury, Ontario and his home was in Owen Sound. Meggason was a farm boy from Goodlands, Manitoba about fifty miles west of Winnipeg. 'Jimmy' Ginn was from Manchester and 'Pete' Campbell was a Londoner who lived in Wembley.

The crew was posted to 1653 Conversion Unit at Chedburgh on Stirlings and then on 10 September came 'the beauty' – Lancaster Finishing School at Feltwell, which was the 'only beautiful aerodrome' Jack Kaiser ever saw: 'three runways in a triangle all about 2,000 ft long in a lovely green grass field with trees outside. In the middle of the triangle was a perfect lake'. When they got closer, in the middle of it was a Lancaster. On the 18th Kaiser's crew was handed over to 3 Group and posted to 90 Squadron at Tuddenham, half way between Newmarket and Bury St. Edmunds.

They still didn't have a rear gunner or an engineer. Very shortly, Sergeant William Clare Telford, a Canadian rear gunner, showed up. He was quiet and never told Kaiser why he was alone so he did not ask him. He had done two trips so the conclusion Jack Kaiser drew was that the rest of his crew had decided they would not continue. 'There were some derogatory names for falling out' said Jack Kaiser 'and though only 20-years old I not only understood how they felt, but sympathized with them. After all, in a tour we were losing one out of every four crews. That is why my squadron was only about 65% manned.

'Flight Sergeant Arthur 'Bill' Wooldridge was assigned to us as flight engineer. He was quite a bit older than I. Even so we got along well right off the bat. I think he was surprised to find I knew quite a bit about our mighty engines. There was no second pilot and it was a mighty machine we were in charge of. It was at Tuddenham that my gunners found out I could shoot as well as them. All Canadian farm kids could.'

Born on 5 October 1913 in Norwich, Bill Wooldridge was one of seven children. He was a good footballer and played for Norwich City 'A' at the old Newmarket Road ground and the 'Nest'. After leaving Quayside School at 14, he got a job maintaining all the leather dressing machinery at a Norwich boot manufacturers. It was there that he met his future wife, Alice Hannent, a copy typist. Alice had spent most of her later childhood in sanatoriums and rest homes with kidney problems. Her father, who was the general manager, may not have been too pleased when she took a fancy to the company 'grease monkey'. However, she got her own way. A romance developed and they married in 1936. Alice was warned that if she conceived it would shorten her life, but she went ahead anyway and their son Michael was born on 1 December 1937. In 1939, while on holiday on the east coast she was rushed to hospital on the assumption that she was expecting again. Sadly, however it was a false alarm and she died of double kidney failure in hospital on 29 August shortly before the outbreak of WW2. Alice was only 25 years old. 'Bill' moved back to his mother's and his unmarried sister took on the role of mother to Michael aged one year and ten months. 'Bill' immediately registered for RAF service as aircrew and was put on the waiting list but was not called up until August 1944, while serving a fireman for the Norfolk Fire Service in Norwich.

'We were assigned to 'B' Flight the third week in September 1944' recalled Jack Kaiser. 'Our flight commander was Squadron Leader A. R. Scott DFC*, a New Zealander.' Born Auckland, 21 November 1919, Scott was a physical instructor who joined the RNZAF in August 1941. "I liked working for him. (Once, in the Mess of course, he taught some of us how to do a Maori war dance. We would do it just before we played rugger). In our flight we were supposed to have 16 crews; we had twelve. As such one got through a tour fairly quickly; that is if one got through. As a 19-year old I "knew" in my little world that it would never happen to me so I ignored it. Also I made it a point not to discuss it with my crew.

'We flew four practice flights before our first operational trip. One was fighter affiliation primarily for gunners/pilot co-operation. Because the manoeuvres were rough we carried only the pilot, engineer and gunners. I really enjoyed these training flights. I could almost bend that Lancaster inside a Spitfire, but not quite!

Jack Kaiser's crew's first op was on 3 October when 252 Lancasters and seven Mosquitoes carried out the first of eight attacks on Walcheren Island, a reclaimed polder, in an attempt to open up the 40-mile River Scheldt entrance to Antwerp. The target for the bombers was the sea wall at Westkapelle, the most western point of Walcheren. "A bunch of Tiger tanks were parked on Westkapelle Island below sea level just outside of Antwerp' says Kaiser. "Our job was to breach the dyke to engulf the tanks in water from the North Sea. We did and they stayed there until after the war. We were inbound at 15,000 ft. It was a beautiful day. We were in the middle of a stream of Lancasters when the Master Bomber called us down to 5,000 to make sure we destroyed that dyke. It was loud and clear but nobody descended at once. I had been told many times to obey the Master Bomber so I got up nerve throttled back, lowered the gear to speed the descent and crossed my fingers. It crossed my mind (that was why I hesitated for a minute or two), that if they bombed from 15,000 and me at 5,000 I had a good chance of swallowing a 4,000 pounder. But they all followed as soon as I lowered the gear and started down. Though it was our first trip I was the first to obey the Master Bomber. When I told the debriefing intelligence officer, all he said was: 'good, we got the damn dyke'.[1] He was a flight lieutenant pilot with a DFC and bar. The next day our picture showed Ross's 4,000 pounder dead centre on the dyke. Our Bombing Leader thought it great. Ross and I thought it normal. You can't beat an old Saskatchewan farmer.[2]

'Our second op was a 5:15 hours flight on the night of 5/6 October. Target: A railway junction just outside Saarbrücken." The raid was carried out at the request of the American Third Army which was advancing in that direction. "We carried nine 1,000 lbs and four 500lbs. Ross of course hit dead centre. His voice was completely 'matter of fact', just as if we were on the range safe and sound in England. After 'Bombs Away' the seconds (about seventeen) seemed like an hour. I had to hold her steady so we would get a picture of our hit. There was little flak. I thought 'pull out of here' and then I thought 'how do I know its any better there, wherever 'there' was from here? It was calm and quiet and I realized that the flak was aimed at the aircraft behind us, so I sat still. Good thing too because the next day we had a perfect picture of our bombs dead centre in the railway junction. The centre of our picture was where our load hit." Three Lancasters were lost.

"On 7 October we flew a 3:40-hour daylight on Kleve, a road junction. We went in at 15,000 ft with a load of eleven 1,000 pounders and four 500 pounders. They were heavy enough to delete a road junction. We were subjected to heavy flak on the bombing run and on the way home at The Hague we got a gut full of 'predicted flak' (flak that searches for a target). I heard the tinkle a number of times but nobody was hit and 'she' (all my aircraft were a 'she') flew home with no evidence of damage on the way. We did not know nor realize how badly 'T' was damaged until we got back to Tuddenham and the aircraft was taken out of service.

"For our fourth op, a daylight raid on Duisburg on 14 October, we were given 'Y-Yorker'. We were outbound in formation but about half way between Antwerp and Duisburg but we were to break formation

---

1. The bombing forced a gap during the fifth wave of the attack and later waves widened the breach until the sea was pouring through a gap estimated to be 100 yards wide. No aircraft were lost.
2. The final raid on Walcheren took place on 30 October and the island fell after a week of fighting by Allied forces who sailed their landing craft through the breaches in the sea walls made earlier by Bomber Command but it was not until after a further 3 weeks before the Scheldt was cleared of mines and the first convoy did not arrive in Antwerp until 28 November.

and make individual bombing runs. Why? I don't know, but I think someone 'up there' felt that if we stayed in formation we would be more concentrated over the target (and thus subject to flak). Or maybe it was to see how we would manage. Whatever, it was a pain in the ass! We had to get up, get fed, be briefed by 0500, and airborne by 0630."

Operation 'Hurricane', as it was called in the Air Ministry directive that Bomber Command had received, was to 'apply within the shortest practical period the maximum effort possible, to demonstrate to the enemy the overwhelming superiority of the Allied Air Forces.' In all, 1,013 aircraft – 519 Lancasters, 474 Halifaxes and 20 Mosquitoes were dispatched.

"We formed up and had to fly for 1:45 hours in formation and then do the bombing run singly and fly back individually' wrote Jack Kaiser. "We had a 'cookie' which was used to knock down big buildings. Our formation position was in the rear so we were nearly last to run in. Ross told me he hit 'the factory' but it was mostly gone by the time we dropped our load. A 'cookie' made a bit of a splash so I asked him if he saw our result. His calm answer was, 'Oh yes we took out the rest of it'. We experienced what I thought was heavy flak on the run in and out and we did get hit. Strangely we went north about forty miles before we turned for home. I quietly hoped that this would not become a regular system; my navigator and bomb aimer were too good for that kind of crap!"

In all, 957 Lancasters and Halifaxes dropped 3,574 tons of HE and 820 tons of incendiaries on Duisburg for the loss of 13 Lancasters and a single Halifax.

'We went back to Duisburg that night for our fifth op' wrote Jack Kaiser. This time 1,005 aircraft – 498 Lancasters, 468 Halifaxes and 39 Mosquitoes were dispatched in two forces, two hours apart. "The 'Doc' gave me two 'Wakey-Wakey pills; 'take one now and one when you leave the target'. I did but I did not notice any difference. We took off at 2215 hours. This operation must have been required because the photo analysis taken by the 'Mossies' that day showed that we did not get the job done. The route was different and we went southwest around London before turning southeast. We crossed over Beachy Head, hit the coast near the mouth of the River Somme and flew on to just south of Brussels, over Düsseldorf and onto Duisburg. We ran in at 22,000 ft. There was little or no flak and the bombing run was like a training exercise. Ross figured he hit the Master Bomber's marker because it went out 17 seconds after we released, exactly at the time our camera took its picture of our hit." All told, 941 bombers dropped 4,040 tons of HE and 500 tons of incendiaries that night. Five Lancasters and two Halifaxes failed to return. "The return was uneventful. With two trips in one day I didn't bother checking the picture the next day rather I slept in to about noon. 'Toosh', our 6 foot, 200 lb cleaning girl brought me a cup of tea at noon. Strangely with all that stress I slept like a top.

'Op No. 6 followed on 19/20 October. It was a 5:55 hour round trip at night on Stuttgart [which involved a total of 565 Lancasters and 18 Mosquitoes in two forces, 4 ½ hours apart]. I did the air test on 'Yorker'. She had a couple of hang ups after being hit by flak on our day trip to Duisburg. All repaired and back in service. I liked that Lanc; its engines seemed to purr when I flew her. We carried a 4,000 pounder and eleven 640 lb SBCs [small bomb containers]. I vaguely remember 'Pete' firing his guns and hollering: 'Fuck, I missed the S.O.B!' The incident lasted about three to five seconds. 'Pete' told us it was a Me 262 twin engine jet fighter that crossed by above us. He figured the pilot may not have seen us. There was only scattered flak so we had a smooth bombing run at 17,000 ft. Of course we did not know what we hit. We bombed on the markers put down by the Master Bomber – we hoped they were in the right place. I preferred night trips because there was less resistance. On our tour I only saw night fighters three times, thank goodness! Our primary opposition was anti aircraft guns.' Serious damaged was caused to the central and eastern districts of Stuttgart and some outlying towns. Six Lancasters failed to return.

Trips to Neuss, a G-H attack, on the 22nd; Essen (twice – a night raid on the 23/24th and a daylight on the 25th) and Cologne (on the 28th) followed. On that second Essen raid 90 Squadron were in the fifth wave of 771 aircraft, 508 of them Lancasters, and bombing was from 21,000 feet. "On "windows" wrote Jack Kaiser "our orders were to start at 6 Degrees inbound and cease at 4 degrees outbound. I drew little windows at each place on my chart, though Bill who dispersed the "Window" by hand out a special chute really needed no guidance from me. He and 'Howie' had it wired between themselves so I need not worry or bother. Like all flights we didn't just climb out to altitude or descend at will. We had special instructions. After takeoff -112 mph IAS (indicated air speed) at 3,000 ft to 03:30 degrees east; climb to and hold 15,000 at 96 mph IAS; at 06 degrees east climb to 20,000 feet, hold 100 mph IAS, The fake target was Cologne which we passed to the left and went straight north to Essen. Our squadron was to do our bombing run at 21,000 ft. We were already using G-H to navigate along with whatever other aids that happened to be available. We usually did the bombing run at 111 mph IAS but supposedly it did not matter because the graticule was somehow hooked to all of the flight instruments and as such took in the correct release point automatically. I was sceptical so I always held the Lanc perfectly steady on the run in. It paid off in that we got the Master Bomber's marker nearly every flight. We stopped checking the photos after 7 or 8 direct hits.

"Somebody in Command got the right idea. Instead of just blundering off home they now kept the "stream" in close by requiring specific altitudes and IAS (Indicated Air Speed) to specific points. Also it was much safer to go home in a bunch than alone. If one got separated, the chances of getting shot down or intercepted increased considerably. I remember discussing and trying to sell that in the Mess one night to a Canadian and a Brit. They did not agree; they got quiet heated and called me a fink (whatever that is), but I would finish my tour; both of them bought it."

"I guess we didn't get it done on the night trip on the 23rd (I was not too surprised because considerable lower cloud made it practically impossible for the Master Bomber to get the Markers in the right place) so we had to go back to Essen in daylight on 25 October and do it right. We carried much the same load, 1 x 4,000, 6 x 1,000, 5 x 500lbs and flew in a five plane 'Vic' formation. We were No. 4 (the outside aeroplane on the right), led by the 'C Flight' Commander Squadron Leader Scott in 'Sugar' with 'Quebec', 'Whiskey', 'Romeo' and 'Yankee' (us) in formation. This was the first time I saw my associates in 'C' Flight fly. They were bloody good and held it right in there. 'Howie' followed the bombing run in, over 10/10ths cloud at 22,000 ft, on G-H. He was not satisfied with the run in. Jim with the H2S and 'Howie' with the G-H told me later that the two units did not agree. This time I was glad I was not carrying the can. 'Scottie' was. My job was to hang in the formation and shut up! I never did find out if we hit anything important. I was glad it wasn't my call, 'Scottie' was the formation leader and when he dropped his we dropped ours. Of course it was better to drop the load in Germany rather than in the North Sea, where we offloaded hang ups and would have had to drop the entire load if not released over the enemy. They were dumped there so they didn't accidentally kill our own armourers after we got home.

"On our 10th op on 28 October on Cologne we went in rather low (19,000) with a load of one 'Cookie' and 12 x 500 lbs clusters. At the briefing the padre had said "Gentlemen, please do not bomb the cathedral if you can so help; it is on the southwest edge of the city." It was standing and looked so peaceful when I saw it, but you can bet I didn't look long. There was particularly heavy flak so Ross got another aiming point; that man was imperturbable. On the back of my map it gave the call sign of the Master Bomber – "Banguo" and Master Fighter – "King Cole" which meant we had some fighter cover though I saw none. Considering the distance in it would have to be US army Mustangs. My friend Don Higgins bought it at Seacheldt."[3]

Jack Kaiser's crew's eleventh bombing operation was another daylight on 30 October, on the industrial city of Wesseling on the Rhine bordering Cologne city on the south when 102 Lancasters of 3 Group carried out a G-H raid on the oil refinery. 'We carried one 'cookie' and ten 500 pounders in 'Y-Yorker' said Kaiser. 'Near the target 'Bill' and I decided to feather the starboard inner engine, which was overheating and was low on oil pressure while we were relatively safe in the stream. Later, when we would be left behind if attacked we could un-feather and use it. Though slower we all agreed it was safer than going home alone. In the bombing run, at 17,000 ft we lost the port inner to flak. Homebound over the Netherlands now on two engines, behind and descending, we lost sight of the stream. We still had the two outer engines so both turrets were operational. Our plan if attacked was to start No.3, dive into a lower layer of cloud at 5,000 ft and hope we could evade them. We had all previously discussed similar possibilities and settled on that general plan. I briefed the crew on the intercom. Howard reminded me of the trip to Wesseling. 'We lost right inner engine inbound, you shut it down when

---

3. Flying Officer Robert Josephus Constable Higgins (23), also known as "Tee Hee" because of his laugh, was shot down and killed piloting Lancaster HK602 on 90 Squadron in the target area at Walcheren. He had married Kathleen Mary Watson, an assistant matron, in the couples' home town of Folkestone in 1943. Only the bomb aimer's body was recovered. Higgins and the five other crewmembers are commemorated on the Runnymede Memorial.

the oil pressure fell off and it started to heat up. We all discussed the best plan and agreed the safest bet was to hang on in the stream rather than return alone. I had to re-rig the 'Gee' to another engine. We were last through the target and got hit somewhat, it sounded like hail on a tin roof. You feathered the left inner engine all we had left was the two outers so we fell behind. You may remember the smell of avgas after we shut down and how we got out of there fast.'

Then Clare reported over the intercom: 'Four fighters behind us and closing fast!' Then: 'It's ok skipper; the fighters are Yankee Mustangs!' They escorted the crippled Lancaster for about twenty minutes until they were out of danger. Jack Kaiser had started No. 3 and then followed the longest two to three minutes of his life until Clare identified them as Mustangs. 'You could hear my sigh of relief in Saskatchewan' said Kaiser. 'They bracketed us and stayed to halfway across the Channel, saluted and left. We landed safely at Tuddenham.'

It was back to Cologne on the night of 31 October/1 November. Then the crews were given a break that lasted to 10 November. Kaiser's crew's 13th 'op' on the 11th was a daylight on Castrop-Rauxel near Dortmund when 122 Lancasters of 3 Group carried out a G-H attack on the synthetic-oil refinery. They flew the op in a new 'Y' as Kaiser recalled: 'My old girl had been put out to pasture when I wasn't there. Engineering told me she was stripped of her engines, props, turrets avionics etc and melted down. Believe it or not I felt sad and I asked the dispatcher not to assign 'Y' to me again if he could help it. Our load was one 'cookie' and 16 x 500 lbs; all there was room for in the bomb bay. We went in over Ipswich, Ostend and north east to Münster and south to Castrop-Rauxel. Our run in was at 19,500 ft. Howard navigated solely by the G-H which performed well. We had no H2S. Supposedly with the G-H we did not need it. However, it meant we had no aft directed radar to pick up a stalking fighter should we get one. All we had for a night trip looking aft for night fighters was Clare's eyes in the rear turret. We got some flak in the left aileron and right fuselage. After a double feint to Cologne we went straight west and inbound the way we went out. 'What the fucking hell were they doing putting a Lancaster on a night flight with no H2S? A visiting Air Commodore, about thirty if that, heard me (as did nearly everybody in the room). We spoke for five minutes. He said he would look into it. Ergo H2S on all Lancs; at least on our squadron and in a matter of days. On 18 November Squadron Leader Scott took us through eight G-H bombing runs in two hours. 'Howie' did pretty good with this highfaluting stuff (average error 60 yards). Even so all three of us and 'Scottie' knew we would never get the accuracy of the Mk.21 bomb sight. The GH's advantage was that we could now be reasonably accurate in cloud or with no markers.

'On 20 November we flew our 14th op – a daylight in 'Y-Yorker' – on Homberg, just south of Duisburg, with G-H now installed. On entry it was the first place I looked. I wonder what I would have done if it was not there. Ross and 'Howie' obviously knew because they were chuckling (quite out of character) all the way from briefing to the pad. We flew the same route out – Ipswich; just north of Ostend to 4 degrees East and 51 North then North East to the Rhine, sharp right to the run in, supposedly Homberg just west of Duisburg. The G-H told us that we bombed the wrong target though we did hit a built-up area. We were above 10/10ths cloud; we turned hard right to our inbound track and followed it home. It was the current route home that missed all the known heavy flak installations." In total, 183 Lancasters of 3 Group made a G-H attack on the oil refinery but the weather was stormy and many crews were unable to maintain formation with the G-H aircraft on the bombing run and the bombing, through cloud, was thought to have been scattered. Five Lancasters failed to return.

After three days' schooling on the G-H equipment at North Luffenham (the main problem was not the equipment but rather the calculation of the release point, thus human error on the ground) Jack Kaiser, his navigator and bomb aimer were considered good enough to use G-H on Neuss on 28/29 November' [when 145 Lancasters of 3 Group and 8 Lancasters of 1 Group carried out a mainly G-H attack.] "There was no flak to speak of" recalls Jack Kaiser "but it was not a good 'prang'. We ran in on G-H and no bomb release point came up so we did not release the bombs. (G-H had its place but not in precise bombing; the Mk 20 graticule on our current bombsight was much more accurate). I tried to contact the Master Bomber with no answer. We could see none behind us and if we didn't bomb here we would have to drop them in the North Sea. I was not about to do so. At 20 years of age I was in this war to win it, not kill a million fish. We were now over the target and supposedly the last wave, so I continued ahead 35 seconds, twice the drop time, turned right on 45 degree bank turn and ran in on the heading for the next short leg. Ross using the bombsight dropped them and figured he had a marker. Unfortunately we very nearly hit another Lancaster inbound on the G-H track. I saw his engine exhaust and pulled up sharply. He went by under us. In retrospect mine was not a good idea, but at the time I thought it OK because we were supposed to be the last wave; obviously we were not! I kept quiet about it at debriefing and no one on the Tuddenham squadrons mentioned it'.

Trips to Bottrop just north of Essen on 30 November, when the target was a coking plant, and two daylight trips, to Dortmund on 3 December, to the Hansa benzol plant, and Oberhausen on 4 December came and went and Kaiser's crew's 19th trip followed on 5 December [when the target for 56 Lancasters of 3 Group was the Schwammenauel Dam on the River Roer]. 'We carried thirteen 1,000 pounders' Kaiser wrote. "It was my opinion at the time that two 'cookie's would be more likely to do the job, but fortunately this time I kept quiet. The philosophy was that a string of bombs was more likely to hit the target than a single and any one of the 1,000 pounders that hit the dam would open up a hole and Mother Nature and gravity would do the rest.

"I liked the call signs of the Master Bomber ('Steel Grey') and Master Fighter ('Tea Cloth') – you can bet he was British. We had to be really careful on this one because the dam was only three miles ahead of our troops if that. If we could attack visually (i.e. bomb sight) the word was 'Fishbone'; if obscured by cloud it was 'Breadbasket' and if we had to abort, because our troops were too close or may even have already taken or destroyed the dam, it was 'Queensbury'. We never opened our bomb doors nor did our formation [the target was cloud covered and only two aircraft bombed]. 'Howie' could get naught on the G-H; there was 10/10ths cloud below us so there was no way we could take out that dam. There was no flak probably because the enemy had their hands full. The Master Bomber said in a very cultured voice 'Gentlemen you get the Marquis of Queensbury; go on home'. We were obliged to jettison three of the thousand pounders off Southwold because of landing weight. We had not jettisoned before. Ross could select any one of those bomb release points or all of them. We did a dummy run on a wave (he said) and all three aircraft in our formation released their 1,000 pounders with us and we went home.

A night trip to Merseburg (Leuna) followed on 6/7 December. "We needed to carry full tanks (1,990 lbs)' recalled Jack Kaiser "so we could only carry 9 x 1000 lbs of bombs or we would exceed T.O. weight. We used the long runway and I watched the end closely, but we got off with about 1,000 ft to spare. I used the B17 T.O. technique. The rate of climb was only about 400 ft per minute. We went out westerly around London and crossed at Beachy Head, along the 50 degree parallel up through the Ardennes well south of Cologne aimed almost directly at Berlin. Fifty nautical miles south west of Magdeburg, we saw a Me 262 (and another was reported, it could have been the same one). For the first time we got a priority on Markers -1st, Red/Green; 2nd, Red; 3rd, Green and 4th, flares. We had a steady run in and Ross was smack on the Red/Green Marker. There was no flak, searchlights nor fighters so the run in was smooth. With no opposition Ross figured and 'Howie' and I were all sure he got the target dead centre. Going home was routine no opposition, flak nor fighters. Even so it was a long hard trip. I, probably the only one, would rather we hit Berlin. Unfortunately, that far in, we seldom got pictures the next day. By the day after we had lost interest.'

On 8 December 163 Lancasters in 3 Group carried out a G-H attack through cloud on the railway yards at Duisburg, No aircraft were lost and there was no opposition but there was flak at the target and Jack Kaiser's crew had considerable damage to the bomb doors and the port inner engine was slightly damaged. "Bill saw it right away, but because it still ran fine I did not shut it down.

Witten followed on 12 December when the target for 140 Lancasters of 3 Group was the Ruhrstahl steelworks. German fighters intercepted the force in the target area and 8 Lancasters were shot down. No bombs fell on the steelworks. Jack Kaiser noted that 'Howie' and Ross did a routine, normal and perfect run in and release. "Having been near the head of the stream we were accompanied (still) by other formations and one or two singles. Ross hit and extinguished a marker flare. Unfortunately, about 20 minutes before the target, near the Rhine, we lost the port outer which supplied power to the 'Gee' and rear turret. We were now vulnerable if attacked by a fighter. The formation moved on without us (as per standing orders) and we were on our own. When I tried to pull Clare out of his turret, he said: "NO WAY – I can still turn this ****** by hand and he stayed there! Bill said: "If we even see a fighter I'll crank No 4 up –

Clare will have power for as long as she runs". We did see two Me109s, but they did not want anything to do with us, thank gosh! We were able to hold on to the end of the stream with maximum continuous power on the three good engines. Near Antwerp I powered back to normal cruise, those engines were too good to beat to death, even though 'Bill', who all the way was imperturbable, assured me they would be OK. They were! Our en route time was about fifteen minutes longer than the rest of our wing.'

'When we got down' said Bill Wooldridge 'they found that the oil regulator 'on' the engine was gone and was only letting a light measure of oil through, instead of going up to 8 or 10 lbs it was allowing only about 2 or 3, it would have eventually ruined the engine if we hadn't stopped it, but that finished up all right.'

Jack Kaiser readily admitted that after landing and shut down he was disturbed 'but, these guys rallied around me, just like my brother used to when I got into a fight playing hockey, which was often!'

'At debriefing Howard did the talking, Ross shut me up. 'Bill' called the engines 'she' so I was not alone. Somewhere near year end I was advised my crew was the most experienced and I was the senior captain on the base. I couldn't have cared less, but 'Pete' figured he had it. He said 'we're gonna get the dirty trips.'

On 15 December 138 Lancasters on Nos. 15 and 622 Squadrons at Mildenhall, 90 and 186 Squadrons at Tuddenham, 75 Squadron RNZAF at Mepal, 115 Squadron at Witchford, 195 Squadron at Wratting Common, 514 Squadron at Waterbeach, 218 Squadron at Chedburgh and 149 Squadron at RAF Methwold on the edge of Thetford Forest were briefed for a raid on the railway yards at Siegen, about 50 miles east of Cologne. Escort would be provided by over 100 Mustangs. The time of attack over the target was scheduled for 1400 hours but the bomber crews were delayed for about an hour before take-off because bad weather was forecast over the fighter stations further east. When the Mustangs finally attempted to take off the fighter pilots were unable to see the end of the runways and their contribution was at an end. So too was the Lancasters, now airborne after forming up, after taking longer than planned. "At about 13:30 hours we were recalled" Kaiser wrote. Nobody was prepared to tell us why the mission was aborted and we were sworn to silence so I could never find out why. We were told not to bring anything back."

Orders were that crews could not land back at their bases until their bomb loads had safely been jettisoned in the sea. Due to the airflow over the detonating pistols fitted in the nose, the "Cookie" would often explode even if dropped in a supposedly "safe" unarmed state and to bring one back was fraught with danger. "There was a special place in the sea where we used to go to release these bombs" says Kaiser "and we were obliged to release our load in the North Sea disposal.[4] When they went off it was like someone giving you a kick up the backside. You could feel it when we dropped them from well over 5,000 ft." Each captain made his own decision about how many bombs to drop in addition to his 'cookie'. It seems that about half the incendiaries carried by the Lancasters were dropped, as well as all of the Blockbusters.

Despite all evidence to the contrary, December 15th 1944 has gone down in history as the day that a UC-64A Norseman with the famous bandleader, Glenn Miller on board was lost and one of the theories is that his aircraft was brought down by bombs jettisoned by a Lancaster returning from the aborted raid on Siegen. (A confirmatory signal to SHAEF on 20 December stated that the Norseman (44-70285) was 'missing and unreported since departure Twinwood Field at 1355 hours).'

A Norseman was reportedly seen going down over the sea by Pilot Officer Fred Shaw, the navigator on Flight Lieutenant Victor Gregory's Lancaster crew on 149 Squadron who had taken off from Methwold at 1137 hours for their first operational flight. However, though a Norseman *was* damaged by bombs jettisoned by Lancasters in the Southern Jettison Area it was not 44-70285.

A formation of seven Stinson L-1 artillery spotter aircraft was being flown from Bexhill to Chartres by glider pilots behind UC-64A Norseman 43-5394 and an Air Sea Rescue Walrus. An article entitled "Airmen Get Through on Wing and a Spray" in the 1 February 1945 edition of *Stars and Stripes* described the 15 December 1944 incident: "434th Troop Carrier Group-Seven glider pilots of Maj. Ralph L. Stream's "Down for Double" squadron thought they had seen most everything but they

---

4. Unlike other No.3 Group squadrons, 149 did not log a specific record of their jettison coordinates for December 15 1944. (*Glenn Miller Declassified* by Dennis M. Spragg). While Kaiser jettisoned his blockbuster bombs in the Northern Jettison Area off Great Yarmouth others dropped their loads in the area off Clacton or the Southern Jettison area in the English Channel.

changed their minds after a recent ferry trip to the Continent. Flying L-1 liaison planes they were 300 feet above the Channel under a heavy cloudbank when suddenly bombs started hurtling down at them through the overcast. The amazed pilots watched huge waterspouts blossom from the exploding bombs, which narrowly missed several of the frail single-engine craft, buffeted them like bouncing balls, and drenched them all with spray. Later the mystery was explained when a break in the overcast showed a flight of Lancasters from which the bombs had been jettisoned, flying high above the cloudbank..."

The formation was almost swamped over the Channel at approximately 600 ft by a series of explosions from bombs jettisoned by Lancasters seen through breaks in the cloud heading for Siegen.[5] The L-1s emerged unscathed and arrived safely at Chartres but UC-64A 43-5394 suffered a landing accident at Villacoublay (A-42) presumably as a result of damage sustained in the bomb explosions. Was this the Norseman Fred Shaw saw? It would explain why the loss of a Norseman was not reported, because 43-5394 simply was not lost. Victor Gregory's crew was not one of the 14 Lancasters on 149 Squadron that flew on the operation to Siegen the following day. Their next op was on 19 December when they were one of the 19 Lancasters on 149 Squadron that raided Trier. On the return leg 29-year-old Squadron Leader John Hamilton Laughlin MBE jettisoned his 4,000lb bomb load in the Southern Jettison Area in the North Sea.[6] Flying Officer Winchester RNZAF aborted because of GH failure and landed 'E-Easy' back at Mildenhall with a hung-up 4,000 pounder in the bomb bay that had failed to release over the target.

Jack Kaiser's crew flew their 23rd operation on 29 December, in "Q-Queenie", to Koblenz, one of the main centres serving the Ardennes battlefront, when 85 Lancasters of 3 Group attacked the Lützel yards north of the city.[7] "Our G-H was U/S but the 'Gee' got us close enough to bomb visually using the bombsight" wrote Jack Kaiser "and as usual the old Farmer (Ross) up front got the marker – again. I would have been

---

5. One Lancaster crewmember recalled that, "The bombs were not designed to detonate on water impact; had they done so, there would not have been a bomb group to speak of, let alone Glenn Miller's aeroplane!" (Methwold History Group).
6. Laughlin's DSO was gazetted 23 March 1945.
7. 162 Halifaxes, 22 Lancasters and 8 Mosquitoes of 4 and 8 Groups attacked the Mosel yards near the main city.

surprised had he not because it was a perfectly calm run in and nobody shot at us.[8] The return home was uneventful. I wrote a sad note on the back of my chart. I quote: "Bert got it on the 23rd while we were away on leave. I was told he was hit flying in the slot and went straight in from 18,000 ft. Nobody got out. He often flew the slot with me. He said he felt safer there."[9]

"And then a mining attack was scheduled for 30 December' wrote Bill Wooldridge. "Each crew at some period during their ops had to do one of these trips. They were low level. We would travel out over the sea. Before each op we had to be careful what we ate. Beans or peas or anything like that would give us wind. Naturally we refrained, but all of a sudden they called the mining attack off, so the next thing we had our meal in the normal way – sausage, chips beans etc – a damn good meal. Then suddenly Kaiser's crew were wanted so we went on the mining attack. Half way home I got the bellyache bad; it was the ruddy beans I ate and they gave me hell and my stomach was pretty rough while it lasted. We went out at about 1,000 ft and could see the waves underneath. When we got over the target we went up to 12,000 ft and dropped our six Mk.6 mines. They were huge, damn great things in big canisters."

"Shortly after we left the mining target" recalled Jack Kaiser "Jim picked up two night fighters below and dead centre behind over half a mile and closing on H2S. I called for NO talk except radio (Jim), Clare (rear gunner) and I. Ross came forward to his turret. We waited as the fighters closed. Fortunately, unlike our night fighters, they had no airborne radar. Clare had them visually and after a few seconds (which seemed like an hour) I heard his breech blocks strike but not fire. (He felt terrible because he had forgotten to put his guns on 'fire' when we crossed the UK coast outbound). Immediately I dove left and down with 75 to 80 degrees of bank and I pulled the control column as hard as I could. We did the shortest, steepest, descending turn I could do. Everybody but me blacked out. Of course I felt the G, and slight grey-out which I had experienced several times before. I rolled the Lanc level at 100 feet

---

8. No aircraft were lost on the two raids.
9. Flying Officer Herbert George Floyd flying HK664/V was shot down on the raid on Trier when 153 Lancasters of 3 Group attempted to attack the railway yards through cloud. All eight crew were killed, the aircraft crashing at Echternach, 17 miles north east of Luxemburg. Only 4 bodies were recovered.

going east. Almost immediately I did a steep turn to the right towards England, started a gentle climb and called Ross: "get up in your turret if they turn left after us or to home we may get a shot at them from below." He jumped up, hit his head on his gun sight, and knocked himself cold. Just then a Ju 88 blasted across our nose about 50 feet above us with about 45 degrees of left bank on. I saw him clearly so I am sure he saw me. I immediately went down to 150–200 feet above sea level, asked Bill for full power and I aimed at England. I never saw the second aircraft and at our low altitude we could not get anything on the H2S. Pete of course never saw them because they were below us and he was above. He did see the second pass, for about one half of a second but his guns were aimed aft, not forward. When we crossed the British coast some turkey shot at us. I had forgotten to put the IFF on Channel D. Our new W/C was pissed off because we had not shot down the Ju 88s!"

On 1/2 January 1945 the Lancasters on 90 Squadron were among 146 aircraft of 3 Group that carried an accurate attack on the railway yards at Vohwinkel without loss but Kaiser's crew's 25th trip was "long and hard'. "My chart was full of various altitudes and IAS we were to fly. We hung in there with it even though it made navigation a nightmare; our track went right between Bonn and Cologne, a bit shaky there. Pete got a shell hole in and out of his mid upper turret which did not hit him.(If you saw how small that turret was and how big Pete was you would swear God had something to do with that – Pete said: "No it was the other guy"). While over the target Pete fired over 100 rounds at a night fighter. We could not confirm his hit because Clare was looking out the back and the rest of us were concentrating on the run in with our eyes on instruments and Ross was looking through the bomb graticule at the ground. The target was well marked, and Ross had visual identification so, of course he hit it. There is no doubt in my mind, Pete shot that fighter down – he called a spade a spade. On the way out of the target we saw a very peculiar thing over the front line where there was huge army activity. A big ball of fire (I know not what it was) obviously predicted on us, as it closed thereto army ground gunners shot it down with light flak. I did not know what it was which added to my concern that I was getting jumpy (maybe an imagined fireball?) Happy New Year!"[10]

---

10. The ball of fire was probably a Lancaster, one of which failed to return from the raid that night.

And then came Nuremberg on 2/3 January, the second longest trip we had [7 hours]" recalled Bill Wooldridge. "At briefing we sat in the second row" wrote Jack Kaiser. "Scottie", who usually sat in front, was missing so Reg Williams and crew sat there. Wing Commander Dunham, our O.C., briefed us. We would be the first in after the Master Bomber and were to hang around to get our interpretation of the success of the raid. So I asked Dunham (he was a nice gentleman) "why us? I thought that was the Master Bomber's job?" He answered, "Master Bombers do not always come back – your bomb aimer has been fully briefed." It was obvious to me our discussion was over so I said: "Yes Sir" and sat down. It was a long hard trip, 'Howie' as usual was almost dead on time and Ross hit the railway yards' markers dead centre. I called 'Hotspur' and advised. To my immense relief he answered: "Right. Go on home." I sure didn't question him; we had always been told to obey the Master Bomber over the target so we left.

On 5 January Kaiser's crew flew a 5 hour, 50 minute trip on Ludwigshafen, their 27th op. "That was where we really knew what it was all about" said Bill Wooldridge. "I got a peculiar impression as I looked out of the window. In front of us out of the nose there was a mass of black spots all over the sky; flak! I felt that I could get out and walk on it at 19,000 ft, a peculiar sensation and I never forgot it. And there were a lot of fighters about, coming in all over the place. We saw some of ours go down; some got direct hits; some of them were making hell of our lot. We had orders to keep straight and not to do any weaving but we did and got told off when we got down but we were not going to stick that ruddy lot. They were around us like flies. However, we bombed and came back; ruddy glad it was all over. It was just about the nastiest trip of the lot."

Op 28 followed for Kaiser's crew on the night of 6/7 January when Neuss was the target. "We bombed with G-H because of weather wrote Kaiser; "it was 10/10ths overcast right across most of continental Europe. My hand written remarks were: "Very good G-H attack, we should do all G-H attack at night. V/Good Run" which meant Howard called a calm, steady bombing run. We, of course, have no idea what we hit and we had long since quit viewing the pictures brought back by the Mosquitoes. I don't know how Ross felt but I just could not be bothered.

On Saturday, 13 January Kaiser's crew flew a 6:20 hour round trip to Saarbrücken for their 29th and penultimate trip. 'We went out without having to go west around London and crossed the French coast just south of Boulogne in a dead straight line" Kaiser wrote. "We bombed eight minutes late behind everybody else because we were obliged to feather No.3. 'Bill' and I were fussy about the engines. Even so we would have started her if we were attacked but there was no opposition. On the way back a large part of England was fogged in and we were diverted to Predannack. It was the only airfield in southern Britain not fogged in. The tower cleared us straight in to land. There were countless aircraft (all Lancs) ahead of and many had headed for home where they could not get in and were diverted. Though last to bomb, we were in the first fifty or so to land. We gave each other room; there was no marshalling, screw-all control and they all parked inches from each other with no accidents. We got a couple of rooms and sleeping bags but a lot slept in their aircraft. There were 300 Lancs parked and no fuel available but we had just enough to get home the next day when I calculated the take off roll on three engines and figured it would be very close so we decided to run up and give No.3 a try. If she acted up Bill was to shut her down. She ran just fine on takeoff – it was weird, almost as if that engine was talking to me – and to 2,000 ft then started to sputter and fart about so we shut her down."

On 15 January Jack Kaiser's crew completed their 30th and final operation when they flew 'P-Peter' and bombed Erkenshwick about five miles northwest of Dortmund. 'On the way back' said Bill Wooldridge, 'Jack talked about shooting up the 'drome, which had been banned for some time because such a big kite might do some damage, but at the finish the skipper decided to have a go at it. He dammed near did everything bar but her upside down. We came down over the canteen and the ground crew said they had never seen anything like it! The skipper got hauled over the coals and was nearly court marshalled but when Jack went in to see Wing Commander Dunham he wanted to know if we wanted to go on Pathfinders! We dammed near scragged Jack. We told him that if he wanted to go he could go his ruddy self, but I could see that 'Pete' Campbell had nearly had it. He was a bundle of nerves. Jack would have gone but we had had enough and we wanted a rest.'

Jack Kaiser recalled, 'Howard supplied an intelligent analysis of the training required and pointed out that the war would probably end before we finished the additional training and we would not finish our tour.

He then voted a clear no! Since his abilities were obviously the key to why we were asked and nobody really wanted to go I told them 'I'd be a rat's ass before I would go without them and voted no!' We still had nothing positive on Scottie so Flight Lieutenant Reg Williams Acting Flight Commander 'C' Flight signed my Log Book. Wing Commander Dunham also signed it.' Wing Commander Peter Francis Dunham DFC (32) who had survived three tours did not return from the second of two successive raids on Wesel, on 19 February, when he was Master Bomber. Leo Richer, a Canadian pilot, was flying on Dunham's left. "We were at 22,000 ft, well over thick cloud cover, when the guns opened up, peppering the sky with their exploding black puffs. My eyes were glued on 'P-Peter', Dunham's Lanc', waiting for his bomb doors to open, when we would do the same. Then it happened. I think I screamed over the intercom to the bomb aimer: "Bomb doors, open!" About two seconds later, BOOM! Dunham and his crew were blown into eternity! We felt the shock wave; we were that close. I think I was in a kind of shock. The deputy commander on Dunham's right took over, and we both dropped our bombs.[11] Dunham's Lancaster crashed in the Rhine near Xanten. He left a widow, Edna Mary Dunham, of St. Albans, Hertfordshire.

'After finishing we were sent home on leave for about six weeks' said Bill Wooldridge. There was a kind of hum drum effect of flying. You had a job to do, you did it and that's where it finished well up to a point, up to the first ten you did not really know what it was all about, on the second ten it just about began to sink in what that meant and on the third ten you were sweating like hell whether you would get through it. All in all, if you were lucky like we were, it wasn't a bad life. You never got too much bull or anything like that. But if you got busted up once or twice it wasn't very pleasant. Once Jack got a piece of flak through his canopy down the side of his face and knocked him groggy for a minute and manage to keep control of the aircraft and asked me to stand by. When I told him he was all right he told me to get out of the ruddy way, so we let it go at that. Then there was another time when Ross got one down the side of his neck. It was nothing much and we shoved a bandage round it. One time we found a hole you could get a penny through that was roughly in line with Howard but we just wrote that off as another bit of good luck. The sum total of my operational life was about 160 hours.'

---

11. From Leo's memoir, *I Flew the Lancaster Bomber*. (lulu.com, 18 September 2013.

Chapter 11

Attacks on The *Tirpitz* 1944

We flew over it, around it, all about it and still it sat there with dignity under a huge mushroom of smoke which plumed up a few thousand feet in the air. There were fires and more explosions on board; a huge gaping hole existed on the port side where a section had been blown out. We had now been flying close around Tirpitz for thirty minutes or so and decided to call it a day, so we headed out towards the mouth of the fjord. Just then Flying Officer Eric Giersch the rear gunner called out, 'I think she is turning over.' I turned back to port to have a look and sure enough she was, so back we went again. This time we flew in at 50 ft and watched with baited breath as Tirpitz heeled over to port, ever so slowly and gracefully. We could see German sailors swimming, diving, jumping and by the time she was over to 85° and subsiding slowly into the water of Tromsø Fjord, there must have been the best part of 60 men on her side as we skimmed over for the last pass. That was the final glimpse we had as we flew out of the fjord and over the North Sea. After a 14-hour flight we landed back at Waddington where the interrogation was conducted by AVM Sir Ralph Cochrane. When asked how it went, my one remark was, 'Well we won't have to go back after this one; Tirpitz is finished.'

Flight Lieutenant Bruce Buckham DFC RAAF, a 26-year-old former Broken Hill Proprietary Company (BHP) Ltd clerk from Sydney, captain of the tour expired crew of a 463 Squadron RAAF movie-Lancaster who were probably the last to see the *Tirpitz* after the attack on 12 November 1944. This crew plus two BCFU [Bomber Command Film Unit] cameramen and 26-year-old Guy Byams of the BBC and W. E. West of Associated Press had flown on the first raid, on 15 September. Now the crew of six Australians and one Englishman and the two BCFU cameramen saw the first wave go in and drop their bombs, getting some near misses and the direct hit amidships. After this only the forward guns continued firing. Then the second wave went in and the Australians saw hits on the stern, amidships and finally on the bows. All the ship's guns stopped firing but the ground defences were still throwing up a screen of light and heavy flak. Heavy black smoke hung over the vessel as *Whoa Bessie*, the movie-Lancaster went down and circled it close in.

At the end of August 1944 orders came from Bomber Command that 5 Group was to examine the chances of attacking the *Tirpitz* at her anchorage in Kåfjorden. If it should be decided that an attack by Lancasters on this remote target, about 1,000 miles from the nearest air base in Britain, was at all possible, the detailed plans were to be prepared by 5 Group Headquarters.

The *Tirpitz* had been attacked several times before. In April 1942 a force of Bomber Command Halifaxes made two successive low-level attacks at night while she lay at Trondheim, but a smoke screen frustrated this attempt. Three months later a Russian submarine successfully attacked her and she had to be kept in dock for six months. Later still the Royal Navy made a most gallant attack with midget submarines; these torpedoed the *Tirpitz*, after penetrating an anti-submarine boom and nets and put her out of action for another six months. The Fleet Air Arm attacked her when she was in Altenfjorden during the spring and summer of 1944; dive bombers from aircraft carriers got direct hits and damaged her. But in the summer of 1944 the *Tirpitz* was repaired and fit to go to sea.

The *Tirpitz* was at this time Germany's last remaining battleship and her mere existence in this northern anchorage kept a considerable British naval force in European waters at a time when every possible warship was needed in the Pacific. More particularly, the presence of the *Tirpitz* in the north was a threat to our sea communications with Russia; every convoy carrying supplies by the Arctic route was liable to attack and the destruction that this capital ship could bring about in a very short time might have amounted to a major victory for the enemy. The *Scharnhorst* had previously made a sortie against one of our convoys to Russia and had only been sunk after a hazardous sea battle; the *Tirpitz* was a much larger vessel, the sister-ship of the *Bismarck* and probably an even more heavily armed and armoured ship. To sink the *Bismarck* after she had been torpedoed by an aircraft from the carrier *Ark Royal* had required an extraordinary concentration of naval power.

Such a warship, which in itself constituted a fleet in being, was naturally well defended by the enemy, but in the anchorage where she then was nature itself gave almost complete protection from air attack. The north Norwegian coast is covered with low stratus cloud which sweeps in off the Gulf Stream when the wind is in any direction except south-east; there are only three days in any average month when the wind may be expected to be in that quarter and on no other day can less than 3/10ths cloud be expected. This circumstance was well known at the time that the operation was being planned; Squadron Leader Peter Furniss DSO DFC of 106 Group had flown many photographic reconnaissance sorties in that part of the world and gave extremely valuable information about the weather there to the 5 Group Meteorological Officer. And a major difficulty was that it was impossible from as far away as Great Britain to predict with any reasonable certainty when the short periods of clear weather were likely to occur, or to predict them sufficiently far ahead to give time for a force to reach so distant a target, especially as aircraft would have to fly first to Scotland and then refuel before crossing the North Sea. During the winter months, to add to the difficulty, there would be no daylight in this high latitude.

But the enemy was not content with natural protection and took every care to guard against the occasional moments when the battleship might be considered in some slight danger of air attack. The Kåfjorden is narrow and steep-sided and the Germans had installed a smoke screen which, as was known from the experience of the Fleet Air Arm in earlier attacks, could fill the whole gulf with impenetrable smoke within ten minutes; it seemed extraordinarily unlikely that an air attack could be delivered without giving the enemy a good deal more than ten minutes' warning, since the Germans had an efficient chain of radar early-warning stations along the coast. Most of the smoke came from drums placed along a road surrounding the anchorage of the *Tirpitz*, but there were also some smoke containers on higher ground to ensure that the smoke would quickly rise above the height of the *Tirpitz*'s mast, which otherwise might be a sufficient indication of her position. In the spring of 1944 some twenty fishing vessels had also been equipped with smoke containers to make sure that the *Tirpitz* would be hidden whatever might be the direction in which the wind was blowing. And there was reason to believe that the enemy had recently become so nervous of air attack, though he might have been expected to think that Allied aircraft would be already sufficiently occupied in supporting the invasion of Europe by sea and by land that the smoke screen was turned on during the summer on every night of good weather.

The *Tirpitz* was very heavily armoured expressly against air attack. Her horizontal armour was in two layers, the upper layers two inches thick and the lower, some twenty feet below the upper, 3.2 inches thick; the armour was even stronger over the gun turrets and magazines. For active defence, besides the *Tirpitz*'s own formidable armament of sixteen heavy and 16 light guns, there were about 38 heavy and 22 light anti-aircraft guns in the Kåfjorden. Fighter protection could be called up in reasonable time and though it might be possible for carrier-borne fighters or Russian aircraft to patrol neighbouring enemy airfields and prevent fighters taking off for the defence of the *Tirpitz*. This would inevitably remove the element of surprise and give the enemy ample time to get his main defence, the smoke screen working.

There was little difficulty in choosing the appropriate weapon with which to attack the *Tirpitz*. Only the 12,030 lb 'Tallboy' deep-penetration, earthquake bomb with 11-second delay developed by Dr. Barnes Wallis could be expected to penetrate her horizontal armour and burst below it; no other bomb in existence would do anything like so much damage to a capital ship. But provision had also to be made for the possibility, or even probability, that the smoke screen would effectively prevent precision bombing with the 'Tallboy' bomb and in this event it was decided to use the bomb, or mine, known by the code word 'Johnny Walker', which had been produced in 1943 for use against capital ships in harbour, though it had not been used because no suitable targets were found; it was only likely to be effective in specially deep harbours. This weapon had no great weight – its weight is variously given as 400- and 500 lb – but large numbers could be carried by a small force. It made frequent ascents and descents below the surface of the sea and the idea was that it would thereby strike the soft under-belly of a capital ship, where there was no armour plate. Large numbers of these mines were to be distributed all over the *Tirpitz*'s anchorage through the smoke screen in the hope of thereby getting a hit. A method of attack was worked out similar to that used by 5 Group against land targets that might be obscured by smoke; the bomb aimers were to get in their sights an easily recognizable lake at the top of the north side of the fjord and by using a 'false wind' setting, by bombing on slightly different headings and by an appropriate spacing of each stick of 'Johnny Walker' bombs, it was hoped to get a bomb pattern about 750 yards square, with the *Tirpitz* in the centre.

The attack would have to be made from extreme range. A Lancaster with a full load of petrol and bombs could not fly to Kåfjorden and get back to this country even from the most northerly airfield in Scotland and any attacking force would therefore have to land in north Russia, either before or after making the attack. No. 30 Mission in Moscow was approached and with the approval of the Russian Government sent particulars of airfields in northern Russia which might be suitable for Lancasters. Yagodnik airfield, on an island in the River Dvina, about twenty miles south of Archangel, was eventually chosen. The airfield was grass covered, with no prepared runways, but the soil there was very sandy and never became boggy, so that it was always fit to take heavy aircraft. But large potholes were apt to appear in the ground and as the Russians were not in the habit of filling these in, heavily laden aircraft might get into difficulties. From Lossiemouth to Kåfjorden and on to Yagodnik was not too far for a Lancaster carrying normal fuel tanks.

A plan of attack was issued in a 5 Group operation order dated 7 September. The code word 'Paravane' was given to the operation. It envisaged that the force would fly to Lossiemouth, land there to refuel, fly to Kåfjorden to attack the *Tirpitz* and then land near Archangel. But the AOC had decided that it was a necessary condition of the operation that there should be forty-eight hours' notice of the probability of suitable weather, which meant no cloud over the target, reasonable weather along the route with moderate winds and good weather both at Lossiemouth and at Archangel; it soon became clear that such conditions could not possibly be predicted as much as two days ahead. On the morning of 11 September the Group meteorological officer told the AOC that there was a spell of unsettled weather ahead in Norway during which there would be only brief periods when the *Tirpitz* would not be covered by cloud. Obviously the only hope of seizing such a fleeting opportunity was to make the attack from as short range as possible and the AOC accordingly decided to send the force to Yagodnik at once; it was to attack from there as soon as the weather was right. This had the further advantage that a force attacking from the east had a much better chance of achieving surprise and beating the smoke screen, since the enemy's radar stations were sited along the coast to give warning of attack from the direction from which it would naturally be expected, that is, from the west.

The force left that evening, between 1856 and 1915 hours double British summer time. It was composed of 38 Lancasters of 617 and 9 Squadrons, 26 carrying 'Tallboy' bombs and twelve 'Johnny Walker' bombs, *"Whoa Bessie"* an RAF Film Unit Lancaster with two camera operators aboard, two Liberators from Transport Command carrying ground staff and equipment and one Photographic Reconnaissance Unit Mosquito, lent to 5 Group by 106 Group, with which to carry out meteorological flights. The force was commanded by Australian-born Group Captain Colin Campbell. McMullen AFC, 617 Squadron was led by Wing Commander James Brian 'Willie' Tait DSO DFC who had lately taken over command of the squadron from Wing Commander Leonard Cheshire and 9 Squadron by Wing Commander James Michael Bazin DFC.

The sudden change of plan on 11 September meant that very hurried preparations had to be made. Thus in the original plan the aircraft carrying ground staff had been detailed to arrive at Yagodnik before the main force and this staff would have been able to get the airfield ready for the arrival of the Lancasters. This was now impossible and in the event this proved a misfortune; it was particularly unfortunate that the ground crews were not there since they could have done much to help the Lancaster crews find Yagodnik.

The long flight was made without incident, except that one Lancaster had to return to base because its 'Tallboy' bomb had slipped, until the force reached the Russo-Finnish frontier, about 200 miles from Yagodnik. Up until then the weather had been as predicted, but from then on it became far worse than had been expected. The forecast had been that the lowest cloud would be at 1,500 ft and that at Archangel, visibility would be six miles. There was 10/10ths cloud with a base of 150 to 800 ft and visibility was sometimes down to 600 yards. The only available map of the Archangel area had insufficient detail for map reading in such bad weather and some of the most important landmarks, such as large towns and railways, were not even marked on this map. The country itself is a waste of marsh, with great pine forests and innumerable small lakes, extremely confusing even if the navigator knew the country intimately and especially so when, as on this occasion, it was necessary because of the low clouds to fly just above the tree tops in heavy rain. Crews had been given the wrong call sign for the Yagodnik ground station but although the force was scattered, all but six of the aircraft landed safely at Yagodnik, Kegostrov, and Vascova and Onega airfields. The Russians gave immediate and efficient help in searching, on 12 September, for the crashed aircraft.

Many of the aircraft which had landed safely were nevertheless unfit to operate without considerable repairs. One needed a change of engine, one had been seriously damaged by flak – there had been some spasmodic firing from Finland – and another had a burnt-out exhaust valve in one engine. Sixteen other Lancasters needed minor adjustments. The engine change was a particularly difficult undertaking for the ground crews; a Lancaster engine had been carried in one of the Liberators, but there was no crane to remove it from this aircraft; accordingly the ground crews built a ramp of tree trunks with blankets on top and slid the engine down this. But even if all the aircraft had been serviceable there could have been no question of attacking the *Tirpitz* that day, that is, on 12 September, because the aircrews were extremely tired. It was estimated that twenty-eight Lancasters, one of them the Film Unit Lancaster, would be serviceable next day, that is if all went well with the business of collecting all serviceable aircraft at Yagodnik as soon as possible, repairing those aircraft which only needed minor adjustments and, of course, refuelling.

But this last did not prove as easy as had been expected. The force had been told they would find eight 3,500-gallon bowsers and four 2,000-gallon bowsers on the airfield, but in fact there were only six bowsers in all, each with a capacity of no more than between 300 and 350 gallons and these small bowsers had of course to be continually refilled. Though refuelling began at once, it took about eighteen hours. This and the work of repair went on throughout 13 September; some aircraft arrived later than had been hoped at Yagodnik and it was not until 0500 hours on 14th September that all serviceable aircraft were refuelled. The ground crews had put in the hardest work of their lives. All of them were very tired and were already feeling the want of sleep when they landed, but they worked continuously for forty-eight hours – and, in fact, until they were ordered away from the aircraft; even then they returned to work after little more than four hours' sleep. This work had been done in great cold and with frequent showers; the aircraft were dispersed along the edge of the airfield for nearly a mile and there was no transport worth mentioning.

As one instance of the complicated repairs that had to be carried out, Lancaster 'W-William' had flown into trees and now needed a new nose perspex, new front turret panel and new engine cowlings and spinners for the port inner and starboard inner engines. Repairs had also to be carried out on the starboard tail plane, starboard bomb-door, starboard wing tip and port undercarriage. With much useful help from Russian technicians, this work was finished on the night of 14/15 September.

There had been more crashes and serious damage among the Lancasters carrying 'Johnny Walker' bombs than those carrying 'Tallboy' bombs and at first it looked as though only five aircraft loaded with 'Johnny Walker' bombs would be ready to operate; however, another was got ready by transferring the load of 'Johnny Walker' bombs from a crashed and unserviceable aircraft to another aircraft which had been forced to jettison its bomb load on the flight to Yagodnik. Even so, this was not enough for there to be any probability of hitting the *Tirpitz*, with this weapon and it was clear that everything would now have to depend on the aircraft loaded with 'Tallboy' bombs and consequently on finding an interval of fine weather and making good use of it before the smoke screen covered the battleship.

On 14 September the serviceable force consisted of twenty Lancasters with 'Tallboy' bombs, six with 'Johnny Walker' bombs and the Film Unit Lancaster; they were prepared to attack that day if the weather should prove favourable. But the Photographic Reconnaissance Unit Mosquito which made a weather reconnaissance in the target area returned with confirmation of a previous forecast of bad weather and the attack was cancelled. The Mosquito made a second weather reconnaissance flight in the early morning of the 15th and returned this time with a favourable report.

The Lancaster crews were immediately briefed. From Yagodnik to the Finnish border they were not to fly above 1,000 ft, but from the border the 'Tallboy' Lancasters were to make a maximum power climb to 2,000 ft above the bombing height and the aircraft loaded with 'Johnny Walker' bombs were to climb to 4,000 ft above bombing height. Three aircraft on 9 Squadron were detailed to serve as wind finders at a height of 16,000 ft. When sixty miles from the target the remaining aircraft were to dive to bombing height; hopefully to beat the smoke screen.

The 'Tallboy' aircraft were to attack in four waves of five aircraft in line abreast at 14,000 to 18,000 ft each wave 1,000 ft in height and with a few hundred yards between each wave. The direction of attack was to be from the south along the fore and aft axis of the *Tirpitz*. The aircraft carrying 'Johnny Walker' bombs were to attack from 10,000 to 12,000 ft in two waves and were to bomb across the target from the south-east to the north-west.

The first aircraft took off at 0630 hours and 23 minutes later the whole force of twenty-eight Lancasters was airborne. The flight to the target proceeded exactly as planned until the final run-up to the target, where there had to be a change of course, because the Lancasters carrying 'Tallboy' bombs were west of track and heading towards Kvaenangenfjorden. The aircraft carrying 'Johnny Walker' bombs were also a little late in reaching the target area.

There was only a little low stratus cloud and this did no more than obscure the target for one or two aircraft on their first bombing run. As they flew in the crews saw 'plumes of white drawing together across the still water and mingling with the soot from the guns'. The smoke screen had begun working about seven minutes before the attack but it was not long enough to hide the *Tirpitz* although it made it very difficult for the crews to judge the result of their bombing. Several crews were convinced that one 'Tallboy' bomb had hit the *Tirpitz* and that five or six had fallen very near to her. Many crews saw a large red flash and then black smoke from about where the *Tirpitz* was anchored, but there was no positive evidence that this had come from the battleship herself.

As to the 'Johnny Walker' bombs it was impossible to see where most of them had fallen, but some were believed to have gone down near the *Tirpitz* and a few fell on land to the east of the target.

Two aircraft were damaged by flak. There was no fighter opposition in the target area, or along the route, either coming or going. Seventeen 'Tallboy' bombs were dropped; two aircraft brought their bombs back to Yagodnik because the bomb-aimers had been unable to see their target and in two other aircraft the 'Tallboy' bombs hung up; it was supposed that the release slips did not work properly because the aircraft had been jolted on the rough surface of the airfield at Yagodnik. After the long flight home all the aircraft, except the Film Unit Lancaster, landed safely at Yagodnik; the Film Unit aircraft returned direct to England and landed safely at Waddington after a flight lasting fourteen hours and thirty minutes.

Two hours after the attack the pilot of the reconnaissance Mosquito made an attempt to photograph the *Tirpitz*, but by then low cloud had covered the fjord again. Flying at 9,000 ft the pilot caught a single glimpse of the battleship, but all that he could say from this was that she was still afloat. It was not until 20 September that the pilot of this Mosquito was able to get a photograph. The *Tirpitz* was then in shadow but the photograph was clear enough to suggest the strong probability that one direct hit with a 'Tallboy' had been obtained. In England it was already known that the attack had been successful and an official communiqué to that effect was issued within a few hours of the operation taking place.

A German report on the attacks on the *Tirpitz* revealed that even more serious damage had been done to the ship than the reconnaissance photograph showed, or than was described at the time in any published or, indeed, in any secret report in Britain. There had been a direct hit on the bows and from the stem to the forward turret they were almost completely destroyed. The main engines were also damaged. It was estimated that repairs, if carried out without interruption, would take at least nine months. A conference on the situation held on 20 September decided that it was no longer possible to make the *Tirpitz* ready for sea or for action again. But she might still have some little fighting efficiency if used to reinforce the defences of the polar area. Accordingly she was to be moved as soon as possible to the area west of Lyngenfjorden, moored in shallow water and used as a floating battery. She was, in effect, out of the war.

When the Lancasters landed again at Yagodnik work was immediately begun on the task of getting the force ready for the return flight to Britain. Only sixteen Lancasters were ready to take off on the next day, 16 September and one of these had to return early with various defects. A further Lancaster was missing from this return flight with then loss of all eleven men on board. The remaining Lancasters took off at intervals in small batches, the last two on 20 September. On 27 September the two Liberators left with the ground crews aboard.

The *Tirpitz*'s move to the west, which had been decided at the conference of German naval authorities on 20 September, did not take place until 15 October. Then a berth was found for her off the small island of Haakoy about four miles west of Tromsø in north Norway and about 200 miles nearer to Great Britain than Kåfjorden, but still beyond the range of a Lancaster fully loaded with fuel and bombs. Shore anti-aircraft guns and smoke-screen units were moved from Kåfjorden to the new anchorage and she was protected against underwater attack and aerial torpedoes by a double net barrage.

On 17 October a detachment of four PR Mosquitoes of 540 Squadron was dispatched to Dyce to keep watch on *Tirpitz*. Information from the Norwegian resistance stared that the ship had left Kåfjorden on its way south for Tromsø Fjord. 'Ultra' code-cipher intercepts revealed that the *Tirpitz* was no longer seaworthy but it was thought possible that the battleship had merely halted at Tromsø on the way to some naval base in Germany where she could be repaired and re-fitted so the War Cabinet decided that the battleship must be sunk. It was not known however, that the German naval staff had decided to use *Tirpitz* as a heavy artillery battery in the defence of northern Norway and her ship's company was reduced from a complement of 2,000 to about 1,600, most of whom were engine room personnel.

By 24 October a plan for another attack on the *Tirpitz* had been formulated. Although no smoke screen had been detected at the new anchorage it had to be assumed that the enemy was getting one ready – and this would involve approaching the target from the landward side; this would increase the return flight from a base in Great Britain to about 2,250 miles. The Lancaster force could refuel in the north of Scotland, but even so the aircraft would have to be fitted with extra tanks. But the bomb load could not be proportionately reduced to take on more petrol, since the aircraft would be carrying 'Tallboy' bombs. The only thing to do was to modify the aircraft and strip them of as much equipment as possible so the ground crews worked day and night shifts. More powerful Merlin engines were fitted and it seemed a justifiable risk to remove the mid-upper and front guns and to remove 2,000 rounds of ammunition from the rear turrets. All portable oxygen bottles and twelve of the fixed oxygen bottles, the Tri-cell flare chute; the pilot's armour plate (if fitted) and all nitrogen bottles were also removed. By these means the all-up weight of each Lancaster was reduced to about 68,200 lb. All aircraft were to carry one Wellington overload tank and one Mosquito drop tank, which would give an extra load of 252 gallons of fuel, or a total capacity of 2,406 gallons of fuel for each aircraft. Even with all these modifications there would be little petrol to spare. Accordingly captains of any aircraft which had less than 900 gallons of petrol in their tanks after the attack, or had engine trouble, they were to go on to Yagodnik airfield, or, if they could not make this base, to Vaenga.

The Lancasters would fly most of the way at a low level, not above 2,000 ft, so the crews would not miss the oxygen supply which had been removed to lighten the aircraft. After bombing at between 13,000 and 16,000 ft the Lancasters were to lose height rapidly to conserve what oxygen supplies were left.

All eighteen Lancasters on 617 Squadron and eighteen on 9 Squadron were to carry one 'Tallboy' bomb each. A Film Unit aircraft on 463 Squadron RAAF would again accompany the force. By 28 October the whole force was ready; 'Willie' Tait was to command and control 617 Squadron and Wing Commander Bazin to command and control 9 Squadron. When in the air, Group Captain McMullen commanded all 5 Group personnel, both air and ground crews, while detached at their advance bases in Scotland and was responsible for dispatching the force when the executive order had been received from 5 Group Headquarters. Some chance of good weather on 29 October was predicted and the force, which now consisted of thirty-seven aircraft including the Film Unit Lancaster, accordingly took off very early that morning. Pouring rain obliterated everything except the lights of the runway. Six hours later, at about 0900 hours, as they came in the crews had a good view of the *Tirpitz* lying at anchor and confidently began their bombing runs. But at the last moment a great sheet of low cloud blew in from the sea and covered the battleship. There was no smoke screen but flak was heavy. Thirty-three aircraft attacked. Though only one near miss by the stern was claimed it caused considerable damage, distorting the propeller shaft and rudder, which flooded the bilges over a 100 foot length of the ship's port side. It meant that the *Tirpitz* was no longer able to steam under her own power. One crew made five runs over the ship in the face of all its defences in the hope of finding a break in the cloud but they had to bring their bombs back. From this operation one aircraft was missing, but this was able to land in Sweden.

Nos. 617 and 9 Squadrons stood by for twelve days at their own airfields. On 10 November they were told to proceed again to Scotland and by noon on the 11th they were there and refuelling began. The preliminary forecast was broken cloud along the route and over the target, with some icing clouds over the mountains that had to be crossed to approach the battleship from inland. But a Mosquito had taken off to carry out a last-minute reconnaissance of the weather in the target area and this returned with a slightly more promising report, though still with no promise of sufficiently good weather to guarantee success; the weather was clear over the target when the Mosquito was there, but there was low stratus cloud about elsewhere and, of course, much might happen in the hours before the Lancaster force could arrive. Thirty Lancasters, each carrying one 'Tallboy' bomb, together with the Film Unit Lancaster, took off at about 0200 hours in the morning of 12 November. As the dawn began to show in the sky the force turned to the east and flew towards the sunrise and the snow-covered mountains of Norway. When there was enough light to see clearly, they turned north again and skimmed over and round the tops of the mountains, keeping close to the ground to avoid for as long as possible being detected by the enemy's radar. There were huge sheets of dazzling white cloud below. Where the clouds broke, the snow now shone a deep blush pink in the light of the red sun; the sky blazed. There was no trace of any living thing in the wild country below.

About 100 miles from Tromsø Wing Commander Tait circled once above a lake and then set course for Tromsø while the other Lancasters moved into battle position behind him. There was no cloud, but Tromsø was not yet in sight; everyone stared in silence ahead. They found Balsfjorden, running almost due south from Tromsø for thirty or forty miles, with steep sides. They were only ten minutes from the *Tirpitz* but a mountain shouldering into the fjord beside Tromsø still hid the place from view. Then, from about twenty miles away, they saw the *Tirpitz*. There was no cloud in the sky, no smoke screen, no fighters.

Suddenly the *Tirpitz* herself opened fire, until she was almost hidden in smoke, pierced by vivid flashes, which rolled from stem to stern; then the guns on the shore and on the flak ships, *Nymph* and *Thetis*, opened fire. The Lancasters were as yet out of range. The bomb aimers gave their pilots preliminary corrections and the Lancasters flew straight towards the battleship, their bomb doors open. They began their final run, each pilot adjusting his speed to that of the leader's aircraft and with anti-aircraft shells now bursting all round them; not a single Lancaster deviated from its course. For thirty seconds before his bomb was released Wing Commander Tait flew entirely by his bomb aimer's instructions, since the target was now beneath the nose of the aircraft. Two seconds before the bomb fell, a red light glowed on his instrument panel. The bomb was released and Tait spun the wheel over to port and dived away out of the flak, back towards the ship but also out to one side of her to see what was happening. The first salvo of bombs had fallen round and on the *Tirpitz* and more were still falling from the last line of aircraft. The battleship was almost entirely hidden in smoke, but as the crew of the leading Lancaster watched they saw a plume of white steam shoot up in a jet through the smoke to a height of 300 ft.

'Either the third or fourth bomb went down into the main magazine beneath the bridge," Buckham said later. 'Then a second hit straight after that. Suddenly, there was a tremendous explosion and the *Tirpitz* appeared to heave herself up out of the water. It was awe-inspiring – a huge mushroom cloud of smoke rising thousands of feet. Two of the big 15-inch guns in the No.1 turret were blown completely out of the vessel into the water and there was fire everywhere. 'We descended to 200 ft and flew over it, around it, all about it but still she sat there. After about 30 minutes, we decided to call it a day and headed back out to the end of the fiord. Just then, rear gunner Eric Giersch called out, 'Skipper, I think she's turning over.' I turned to port to take a look and, sure enough, she was. 'So back we went. I swept in at about 50 ft above the water and we could see the German ratings, 40 or 50 of them, jumping into the water. 'Then we watched with bated breath as the *Tirpitz* heeled over on to her side, ever so slowly and gracefully.'[12]

Two hours later a reconnaissance Mosquito was over Tromsø and the pilot returned to report that the *Tirpitz* had capsized. She was lying in shallow water, with her upper works resting on the bottom of the sea. Two direct hits by bombs of heaviest calibre were certain and several

---

12. Buckham was awarded the DSO. Navigator, Flight Lieutenant R. W. "Doc" Board received a bar to his DFC Cross and four other crew members received the DFC. Wing Commander 'Tirpitz' Tait was awarded the third bar to his DSO.

near misses were observed. The effect of the bombs could be estimated from the bomb craters on the island of Haakoy: the bombs had fallen on massive rocks and the craters had a diameter of 98.4 ft and a depth of 32.8 ft.

At least two 'Tallboys' hit the ship, which capsized to remain bottom upwards. Around 900 men were killed, drowned or suffocated, having been trapped on board in watertight compartments. Only 87 sailors were recovered from the ship by cutting through the double bottom from the outside. One Lancaster, on 9 Squadron, was severely damaged by flak and landed safely in Sweden with its crew unhurt.

The final attack and the sinking of the *Tirpitz* was of the greatest strategic importance in releasing British naval power from its watch on a potential menace to allied convoys to Russia and in permitting a large transfer of forces to the Far East. Once again it had been shown that a bomber could sink even the most heavily armoured battleship.

After the Second World War the Lancaster continued to be operated in significant numbers with the RAF until the introduction of the new Avro Lincoln derivative of the Lancaster. In December 1953, the final Lancaster in service with Bomber Command was pensioned off. In late 1954, the last Lancaster in active service with the RAF, an aircraft which had been used for aerial reconnaissance, is believed to have been retired.

One big Lancaster 'Hurrah' had taken place in 1946. On 1 November 1945, after the defeat of Nazi Germany and Japan, Clement Atlee the British Prime Minister appointed a committee under the chairmanship of the Home Secretary, James Chuter Ede to formulate plans for official British Commonwealth, Empire and Allied victory celebrations. The celebrations took place in London on 8/9 June 1946 and consisted mainly of a military parade through the capital and a night time fireworks display. Most of Britain's allies took part in the parade, including representatives from Australia and New Zealand, the United States, France, Belgium, Brazil, China, Czechoslovakia, Denmark, Egypt, Ethiopia, Greece, Iran, Iraq, Luxembourg, Mexico, Nepal, Netherlands, Norway and Transjordan. The parade arrangements caused a controversy surrounding the lack of representation of Polish forces. The only other allied countries not represented at the parade were the USSR and Yugoslavia. The first part of the parade was the Chiefs of Staff's procession, featuring the British Chiefs of Staff together with the Supreme Allied Commanders.

This was followed by a mechanised column which went from Regent's Park to Tower Hill to The Mall and then back to Regent's Park. It was more than four miles long and contained more than 500 vehicles from the Royal Navy, the Royal Air Force, British civilian services and the British Army. Next came a marching column, which went from Marble Arch to The Mall to Hyde Park Corner. Next came units of the navies, air forces, civilian services and armies of the nations of the British Empire, followed by units from the Royal Navy and then British civilian services, the British Army, representatives of certain Allied air forces and the Royal Air Force. This was followed by a fly-past of 300 aircraft, led by Douglas Bader.

After sunset, the principal buildings of London were lit by floodlights, and crowds thronged the banks of the Thames and Westminster Bridge to watch King George VI and his family proceed down the river in the Royal barge. The planned festivities ended with a fireworks display over Central London. However, crowds continued to gather in London and surrounded Buckingham Palace even after the Royal family had retired from the festivities. Many festival goers could not return home that night and spent the rest of the night in public parks and other public areas around London.

"Volunteering to pilot a Lancaster on Victory night" reported the *Walthamstow Guardian*, "a young airman from Walthamstow thrilled merry-making crowds when he few above the West End after a flight of nine Lancasters, planned to synchronize with a searchlight display, had been abandoned because of adverse weather conditions. He was Flight Lieutenant Dennis James Lundy DFC, who has served in the Air Force for five years and piloted Lancasters on 31 bombing 'ops' and is due for demobilization in a few weeks. Saturday's flight was his last in the RAF. It had been planned that the Lancasters should be caught in the beams of the searchlights just before 11pm. When bad flying weather grounded the aircraft, Lundy went up from his Camberley station with Richard Dimbleby of the BBC as his passenger. Dimbleby broadcast from the air a description of the West End crowds. Earlier in the day, the Walthamstow pilot flew over the Victory procession, taking with him Raymond Glendenning, the radio commentator, two press photographers and a Reuter's reporter. Glendenning gave two short commentaries from the aircraft. 'London looked marvellous from the air' Lundy told the

Guardian after the parade, when he arrived home for Whitsun leave. 'We must have had the best view of the parade. We saw the King and Queen leave Buckingham Palace and could pick out the people on the saluting base. The weather was alright in the morning and we flew between 500 and 800 ft above the parade. During the evening, however, we were in cloud for 50% of the time. We were in the air for two hours and we saw the floodlit buildings and some of the fireworks. It was just wonderful over London, with all the lights on. We took Raymond Glendenning to an air station in Lincolnshire and all the way northwards we saw bonfires blazing.'

Meanwhile, in the immediate post-war period the modification of abundant military aircraft like the Lancaster into desperately needed civil transports was common in the United Kingdom and also, a series of Lancaster navigation flights of epic proportions would be flown around the world.

Chapter 12

Aries

Late in 1944 when the war in the Pacific was still raging, Britain felt that RAF Bomber Command should be used to help America defeat the Japanese Empire and bring an end to WWII. Flying long missions in Asia would entail very long distance flights in extreme climates and accurate navigation would be paramount if long-range RAF bombing was to succeed. In March 1944 when the Central Navigation School moved to the Empire Air Navigational School Shawbury, Shropshire a Lancaster I (PD328) was among the new aircraft on hand and would be among the aircraft used to glean the information needed before bombing operations could begin. PD328's squadron letter was 'A' and as the School's practice was to name their aircraft after stars, planets and constellations, the Lancaster was christened 'Aries'. The Pacific war would end with the Japanese surrender in September 1945 before 'Tiger Force' could be fully mobilised and the Very Long Range Bomber Force was officially disbanded on 31 October 1945. But there was still a need to develop British long range navigation to all parts of the Empire and beyond. Early in 1945 two liaison flights, one to the USA and Canada and one to South Africa, were followed in May by one of the most spectacular flights ever made – to the North Poles (geographic and magnetic).

One hot summer's afternoon in 1945 Wing Commander Bill Anderson OBE DFC had two hours to wait at Newton Abbot railway station. Anderson was a schoolmaster in peacetime and joined the RAF in 1940 as a pilot officer in administration. The death of an air crew friend led him to volunteer for flying and, despite 'advancing age' and a suspect right eye, he managed to pass on to training in Canada and then on to bomber operations on 9 Squadron at Honington. He soon discovered another inherent 'weakness' – he was air-sick every time he flew. He went on to make a name for himself as a wing commander and a Master Bomber in the Path Finder Force.[1] 'I browsed round the bookstall and bought myself a little book. It was entitled *Etiquette*. It told me everything that I

---
1. *Pathfinders* by Wing Commander Bill Anderson OBE DFC AFC (Jarrolds London, 1946).

ought to know. How to write to an earl, how to arrange for an introduction to a lady at a dance, and best of all, a complete chapter on 'How to propose'. 'Be firm, yet not arrogant. Remember that you have a right to expect that your proposal will be listened to with attention.' How I longed to show that to the Commanding Officer of the Training Unit for I was going to him with a proposal, which sounded so peculiar. Would he 'Listen with attention' and 'Incline the head gently to indicate assent' or 'Shake the head with an expression of deep regret'? Or would he just bellow 'Out!'

'I could almost see the look on his face when I made my ridiculous request, 'Please, sir, may I go to the North Pole?'

In May 1944 a flight to the North Pole was not possible and the first experimental long range navigation flights were made in Stirlings; to Ireland (which was soon ended when the aircraft crashed on takeoff from Shawbury after the pilot, Wing Commander David Cecil McKinley DFC had ordered the crew to bail out) and, on 2 June, to Dorval (Montreal) via Prestwick, Reykjavik and Goose Bay. This trip out and back proved successful. In September No.3 Specialist Navigation Course successfully flew three Stirlings to Canada and return. Lancaster I PD328 meanwhile, would be converted to undertake a round-the-world navigational flight with the installation of a ton of the latest navigational equipment including 14 aerials along the fuselage and so the 'Aries Project' to collect data on the aircraft in extreme climates and to study American methods of navigation in the Pacific theatre of war was born.

The Lancaster was named after Lancaster, Lancashire; a Lancastrian is an inhabitant of Lancashire. The Avro 691 Lancastrian was basically a modified Lancaster bomber without armour or armament and with the gun turrets replaced by streamlined metal fairings, including a new nose section, the type saw service as a Canadian and British passenger and mail transport aircraft of the 1940s and 1950s. The initial batch was converted directly from Lancasters; later batches were new builds.[2]

---

2. Nine Lancaster XPP aircraft were built by converting Lancasters at Victory Aircraft Ltd Canada. Specifications: Lancastrian C.1; A total of 23 nine-seat transport aircraft for BOAC and QANTAS and the RAF (designation Lancastrian C.1 to Specification 16/44) were built by Avro. A total of 33 Lancastrian C.2 nine-seat military transport aircraft for the RAF were built by Avro. A total of 18 Lancastrian 3 13-seat transport aircraft for British South American Airways were built by Avro. Eight Lancastrian C.4 10 to 13-seat military transport aircraft were built by Avro for the RAF. The Aviation Safety Network, part of the Flight Safety Foundation, records 23 hull loss accidents involving the Lancastrian occurring between 1946 and 1964.

In 1943, Canada's Victory Aircraft converted a Lancaster X bomber for civil transport duties with Trans-Canada Airlines (TCA). (After the war Victory Aircraft was purchased by what became Avro Canada). This conversion was a success resulting in eight additional Lancaster Xs being converted. The 'specials' were powered by Packard-built Merlin 38 engines and featured a lengthened, streamlined nose and tail cone. Range was increased by two 400 gallon Lancaster long-range fuel tanks fitted as standard in the bomb bay. These Lancastrians were used by TCA on its Montreal-Prestwick route.

Lancastrians also served with the RAF and were used during the Berlin Airlift to transport petrol; fifteen aircraft made over 5,000 trips. In 1946 a Lancastrian operated by British South American Airways Corporation (BSAA) was the first aircraft to make a scheduled flight from the then-newly opened London Heathrow Airport. With the advent of gas turbine engines there emerged a need to test the new engines in a controlled flight environment in well instrumented installations. An ideal candidate emerged as the Avro Lancastrian which could easily accommodate the test instrumentation as well as fly on the power of two piston engines if required. Several Lancastrians were allocated for engine test-bed work with turbojet engines replacing the outer Merlin engines or test piston engines in the inner nacelles. Fuel arrangements varied but could include Kerosene jet fuel in outer wing tanks or fuselage tanks, with AVGAS carried in remaining fuel tanks.

On 20 October 1944, in the expert hands of Wing Commander McKinley, *Aries* left for Australia and New Zealand, returning to Shawbury on 14 December having established 'a practical liaison between the Empire Central Navigation School training and operational units under RAAF and RNZAF control'. This epic fight entailed a distance of 36,000nm over 20 stages and a total flying time 202 hours, of which 15 hours 8 minutes was the longest non-stop stage. The tour lasted 53 days of which 40 lectures were given. McKinley was awarded an Air Force Cross for the mission. In February 1945 McKinley and *Aries* made a 14,595 round trip to North America. In March a 13,977 mile round trip was flown to South Africa and Palestine by Squadron Leader Imrie.

Aries had not differed externally from other Lancasters except for an astrodome in place of a mid-turret, but at Waddington in April 1945 a transformation took place, to modify the aircraft for flights to the North Geographic and Magnetic Poles.

Wing Commander Anderson had eventually ignored the advice of the book on etiquette and took the risk of 'entrusting a thing as delicate as a proposal to paper'. "So I never saw the Commanding Officer of the Training Unit's immediate reaction to my suggestion. All that I know is that a couple of days later a formal request for the flight to take place was forwarded to Air Ministry, together with a suggestion that two pupils on the 'Archangel' Navigation Course should be allowed to navigate the aircraft. The pupils were 'Flash' [Flight Lieutenant S. T. Underwood, navigator/plotter, responsible for maintaining a continuous record of the aircraft's position] and I.

'I have often been asked the obvious question, 'Why did you want to fly over the North Pole?' Luckily it has always been easy to answer this, because this was the first thing that the Air Ministry wanted to know.

'First of all, the Royal Air Force has to be ready for emergencies. It is always possible that a flight over Polar Regions may one day be necessary, for a glance at a schoolroom globe of the world will show that the North Pole is the centre of civilization. And therefore it was obviously wise to find out exactly how such a trip could be carried out and if any special equipment would have to be installed in the aircraft. It was hoped that the flight could be managed only using gear which every aircraft that carried a navigator would be bound to have on board. The first object of the flight, therefore, was to see whether journeys over Polar Regions could be carried out with standard Air Force navigation equipment. At the same time we hoped to discover how the various radio and radar aids would work. Would our 'magic eye' be satisfactory over snow or pack ice and would our other radar aids tend to be unreliable? More important even than this, we wanted to know how the various sorts of magnetic compasses that were in service would work near the Magnetic Pole. In other words, in addition to the navigation, we hoped to carry out a good deal of research.[3]

---

3. The *Aries* expedition was given several tasks – to test McClure's theories of grid system navigation, search for the North Magnetic Pole, identify problems associated with Arctic flying, assess performance of existing instruments, collect weather data, investigate possible radar mapping of ice fields, observe effects of polar flying on aircrew and collect engine and airframe data. Nailing down the position of the Magnetic Pole was important because its position had shifted since it was last discovered.

'Finally, it was a rule on the 'archangels' navigation course that pupils had to carry out at least one very long flight. 'These were arranged to India or South Africa. We asked instead to be allowed to fly in Polar Regions, for the navigation up there would be infinitely more difficult and so much the more interesting. To give an idea of the problems, imagine yourself in an aircraft arriving at the North Pole. The pilot asks you which way he must fly. What are you to tell him? For once at the North Pole, the only direction in which you can fly is south and that is in every direction at the same time!

'The Air Ministry considered our proposals, added a few more ideas of their own and the scheme was duly approved. So 'Flash' and I got down to the problem of the navigation. 'Flash', named after a famous newspaper strip cartoon, is a curiously deceptive individual. He always looks a little tired. One day it was suggested that he might play football in the station trial game. As most people expected, he was not much good, so that instead of playing for the station that Saturday, he had to go away and play for a first-class professional soccer team before a crowd of many thousands!

'For many hours 'Flash' and I worked together, helped by 'Arnold' [Squadron Leader A. J. Hagger] who was the Chief Instructor on our course and incidentally was to fly with us as second pilot and meteorological observer and photographer. After a good deal of hard work, we felt that the time had come to put our theories into practice. I remember well the first attempt. It was most exciting.

'Arnold,' his eyes glued to the instruments before him, was fighting to keep the aircraft on a level keel. 'Flash' was working desperately at the plotting table, checking and cross-checking his work. I was in the astrodome shooting with the sextant, oblivious to everything but balancing the bubble in the eyepiece.

'The tension rose. All our past training, all our carefully worked out theories were now to be put to the test. Time stood still. Then quite suddenly as I laid down my sextant, 'Flash's' voice came through the intercom, 'We are at the North Pole – Now!'

'Somebody switched off the motor. The vast dome studded with twinkling stars that was slowly revolving over our heads driven by a motor at the exact speed to allow for the rotation of the earth came to a stop. We climbed out of the fuselage of the astronomical trainer and went downstairs to look at the track which an ingenious recorder had marked automatically on a special chart. We had arrived at the North Pole – in the Central Link Training School at Elstree near London!

'This was a very important part of our training for the real thing. It enabled us to test our methods of navigation as far as the Geographical Pole was concerned. For the Magnetic Pole we had another problem. The Magnetic Pole is simply a place where the magnetic forces of the earth pull straight downwards. This place or places, for there are probably several, is many hundreds of miles away from the Geographical Pole. To prepare ourselves for this area, we used to fly round England without using our compasses at all, relying simply on the sun and moon to give us directions. And meanwhile, on the ground others were even busier than we were.

'Under the eagle eye of 'Mac' [Wing Commander David McKinley] the skipper, with the help of Avro's and Bomber Command, the aircraft was stripped of turrets and cleaned of paint, for that meant extra speed. The bomb bays and the long pointed nose were filled with petrol tanks and extra banks of oxygen cylinders were fitted.[4] RAF Farnborough installed a vast array of formidable compasses.[5] Wing Commander McClure, our scientific observer, collected a number of remarkable instruments with which he hoped to test the mathematical and magnetic theories which the Astronomer Royal had propounded. The MO [Wing Commander Dr. Roland Winfield DFC] accumulated with a certain morbid relish concentrated food against unhappy landings on the ice. And sundry sportsmen in Cambridge endured the rigours of refrigeration to test the special Arctic clothing which we were to carry."

Yet it was impossible to solve all potential problems. *Aries* had been designed as a bomber and that imposed limitations. Space was limited and crew comforts could only partly be accommodated. Wing Commander Winfield,[6] medical officer of the expedition, wrote of some difficulties:

---

4. The fore and after turrets were replaced by fairings, extra tankage was placed in the bomb bay, increasing maximum range to 5,000 miles or about 21 hours continuous flight and a Lincoln type undercarriage was fitted. A silver polished skin replaced the original camouflage.
5. Eleven different types of magnetic compasses were distributed throughout the aircraft and special recording and navigational equipment was installed.
6. See *The Sky Belongs To Them* published by William Kimber in 1976. Dr. Winfield BA MBChB died at Cambridge on 1 November 1970 at the age of 59. He left a widow, Christine.

'During the flights, one of the major problems was to keep warm the observers working in the fuselage. In the standard Lancaster, the crew are placed either in the cockpit towards the nose of the aircraft or in the gun turrets. Those in the cockpit are warmed by an efficient hot-air system of cabin heating, while the gunners wear electrically heated suits. The fuselage is uninhabited and heating is not required. When the *Aries* was converted from heavy bomber to flying laboratory, it was not possible to warm the fuselage; and since the generators were already loaded to their full capacity to provide current for the scientific equipment, the observers in the fuselage could not use electrically heated clothing. In fact, it became so cold that after spending an hour in the fuselage, it was necessary to go forward to the cockpit for about five minutes in order to thaw.'

'All this impressive paraphernalia meant, of course, extra weight, so that a rather stouter undercarriage had to be fitted and four new Rolls Merlins installed. And these alterations were not made just by waving a magic wand. They represented many hours of real hard work for the ground crews. It was their efforts that made the flights possible.

'VE Day [8 May] came just when the last final modifications were being carried out, so 'Mac' and the boys worked steadily through the holidays! 'Mac' is a delightful mixture of Irish charm and Scots shrewdness. He had just returned from a trip round the world, a record-breaking feat for which he had been cheered and feted in no small way. And at intervals on the polar flight he was to be cheered and feted, too. Yet the times I shall remember him best were after a long and hard day's flying when the aircraft had landed and we would collapse into bed. 'Mac' would sneak down to the aircraft where the ground crew would be overhauling the engines. There you would find him with his shirt-sleeves rolled up, changing plugs far into the night.

'At the same time as we were learning to navigate and 'Mac' was getting the aircraft ready, 'Ken' [Kenneth Cecil McClure], a large, easy-going Canadian navigator, whose name is known all over the world for his work on the special system of polar navigation which we used, was collecting research equipment.[7] He was supervising the installation of a vast array

---

7. David Cecil McKinley of the RAF and Kenneth Cecil McClure of the RCAF had been studying the problems of navigation in Arctic regions since 1942. McClure in particular had worked out a method he dubbed the 'Greenwich Grid System' by which all headings

of formidable magnetic compasses and collecting a number of remarkable instruments with which he hoped to test the magnetic theories which the Astronomer Royal had propounded. His assistant was 'Doc', whose job amongst others was to examine the medical effects of the long, hard flights. 'Doc' had tried everything else in flying except polar trips! He had been parachuted into enemy country; he had flown over a hundred different and dangerous sorts of operations. A few days before the polar trip he was acting as a live dummy being snatched off the ground by an aircraft! Yet, of course, he was only doing all these fantastic things for purely medical interest! 'Doc' was in charge, naturally, of the feeding and he accumulated with a certain morbid relish concentrated food against unhappy landings on the ice. He also arranged our clothing. Sundry sportsmen in Cambridge had endured the rigours of refrigeration to test the special Arctic clothing that we were to use. The most flamboyant article of attire was the string-bag vests that we were expected to wear next to the skin! However, apart from this, the clothing was pretty normal.

'Preparations included mock flights in a device called the Celestial Link Trainer – the most advanced flight simulator of the day. Existing records of Arctic navigation and flying were studied. Winfield sought out suitable clothing, food and survival equipment; clothing in particular had to allow freedom of movement in a cramped aircraft, had to be cool enough to avoid overheating, yet give protection in the event of a parachute descent or emergency encampment awaiting rescue. Prior to departure the crew lived two days in a refrigerated chamber eating dehydrated meals.

'Soon all the preparations were made and we were ready to go. Everything that could be done for our safety was done. It was even rumoured that husky dogs were being trained to be dropped by parachute in case we force-landed on the snow and provisional lists of priority for the members of the crew to be used as husky meat were flippantly prepared! So on 10 May 1945, two days after VE Day, the Lancaster *Aries* set out for Meeks

---

and readings were calculated relative to the prime meridian. His interest in the subject was appropriate; in the 1850s his ancestor, Sir Robert John Le Mesurier McClure (28th January 1807–17th October 1873) had been engaged in the Franklin search which also traced the elusive Northwest Passage. In 1854, he was the first to transit the Northwest Passage (by boat and sledge), as well as the first to circumnavigate the Americas.

Field, Iceland, all ready for the big trip, exactly a hundred years after Sir Benjamin Franklin had set out to cover very similar ground in his ship. But Sir Benjamin Franklin did not return from his sortie.

'There were eight air crew: two pilots, two navigators [Anderson, the senior navigator, responsible for external observations needed to plot position such as astral bodies, and Flight Lieutenant S. T. Underwood, navigator/plotter (responsible for maintaining a continuous record of the aircraft's position); Wing Commander McClure the senior observer (responsible for collecting magnetic, radar and other special data outside of navigational observations and the co-ordinator of research); the MO (and assistant of McClure) and two wireless operators (Flying Officer S. Blakeley and Warrant Officer A.S. Smith)]. In addition we carried three ground crew members to help service the aircraft.[8]

"We flew out to Iceland to carry out a short 'shake-down' test flight up the Greenland coast. Iceland is a bleak place, but nothing like so cold as one expects. In the early days of the war there had been a considerable German population. Indeed, when our ships first sailed into Reykjavik, the Germans all turned out and lined the quay, shouting 'Heil Hitler', for they thought we had come from their Fatherland, cleverly disguised as Englanders. It is believed that this welcome gave the authorities a very good view of Iceland. Base rumour had it that they at once presented the Icelanders with a lovely, big, black railway engine. A lovely, big, white elephant would have been a little more useful for the island has no railways. And to complete the libellous tale it is only necessary to add that the Icelanders gave the engine back to the Air Force who put it on a camp and used it as a model steam laundry! The inhabitants, naturally, would not have thought of this, as they get all their hot water automatically from hot springs!

'We carried out a short nine-hours' 'shake-down' flight up the Greenland coast to 'Little Pendulum Island' to make sure that everything was working properly and then sat down to wait for the weather to clear.

'In the chilly dawn of 16 May we took off for the Geographical Pole. Soon we had left Iceland behind and were setting out for the Greenland coast. It was the last land we were to see for eight hours, for we ran into thick cloud. We hunted for gaps, but it seemed to be solid all through and

---

8. Corporal W. S. Gardner, airframe mechanic; Leading Aircraftman E. M. Wiggins, aero-engine mechanic; Leading Aircraftman H. J. B. Dean, electrician.

we could find no lanes. We tried to fly through it, but ice was forming on the wings and the tailplane and the fins. At last, after four or five hours' trying to sneak through, our engines began to lose power, so we turned wretchedly back to Iceland. We landed after nine hours' flying, having accomplished exactly nothing.

'Two hours later we were airborne again, setting course in a north-easterly direction, hoping to sneak round the back of the weather while it was not looking. Once more we ran into cloud. Once more the edges of the wings and tail began to collect ice and the wireless aerial and insulators glittered white. But the cloud was not so thick. In places we managed to find fairly clear lanes and the engines still developed full power. At last, after seven hours' struggling amongst the cloud, we broke out into glorious clear weather on the north-east coast of Greenland.

'Black mountains, separated by huge sheets of pure white snow, rose up out of the crazy pavement of the pack ice of the Arctic Ocean. Incredibly jagged heaps of rock split by deep gorges down which the glaciers ran to the frozen sea. A fearful and wonderful picture.

'Soon Greenland was left behind and we set out over the sea to the North Pole itself. The ice was broken up into large round slabs, like giant water-lily leaves, with smaller slabs between. But as we went on, the ice appeared to close up and became a sheet of white, split here and there by curiously straight lanes of water running criss-cross over the surface.

'Soon low cloud began to cover the sea, but by then we were too busy to spend much time looking out of the window. Wisps of mares' tails appeared above us and all around the horizon we could see the deep blue Arctic haze. It became very cold in the back of the aircraft. The windows frosted up and once, when some tea was spilt, it froze as it ran along the floor. The next two hours were hours of very hard work, 'shooting' the sun and the moon.

'We had often thought of what we would do when we reached the Pole itself, but when we did arrive there was less celebration than one might expect. I think we were all pretty tired and we had used up rather more than half our petrol and we did not know what the weather would be like on the way back. I had intended to write a letter to Mary for I had so often been on top of the world with her. But when it came to it, all I could do was to get out the pencil and paper and just sit and look at it. However, the wireless operator was at the moment asking Iceland what his direction was from their station and to his great delight they answered north exactly! The skipper did a smart circuit in eighty seconds – 'round the world in eighty seconds, not days!' – passing from Wednesday back to Tuesday and then to Wednesday again in less than half a minute. Meanwhile 'Doc' threw out a large Union Jack and Warrant Officer [A. F.] Smith, the second wireless operator, solemnly chewed a banana – 'first man to eat a banana over the North Pole.'

'The journey back to Iceland was a long, hard haul. There was plenty of cloud both above and below; indeed we ran into it soon after leaving the Pole. We set course to the east to avoid the weather and, to our surprise, we could see open water below us even while we were well north of Greenland. There seemed to be quite large patches and in these areas of open sea, a few little ice floes would be clustered together as if they were holding a regatta. Soon the low cloud thickened and we were in bad weather once more. But this time the icing was much milder and gave us no trouble. So after nineteen and a half hours' flying we landed back in Iceland, pretty tired after our last thirty hours' work, during which we must have covered about six thousand miles.

'Next day we set out for the Magnetic Pole, or rather for the point in the Booth Peninsula at the north-western corner of Hudson Bay, which on our chart was labelled rather naively: 'North Magnetic Pole'. For as we have said, the Magnetic Pole is nothing more than an area where the magnetic forces of the earth happen to pull straight downwards. Indeed there may be several of these areas at the same time and to complicate matters they move about with the time of day and also with the passing of the years.

'We left Iceland in the early hours of the morning and climbed up to cross the Greenland coast. Cloud once more covered the ground below, until we were half-way across. Then it cleared and we saw beneath us a vast, white plain stippled by the wind. It is believed that Greenland is a land of mountains, but that centuries of falling snow have packed the valleys and drowned all but the high peaks around the edge, so that the middle appears as a great plateau. We flew on until the mountains on the far coast began to appear and the sheet of snow began to divide into glaciers, slipping down the sides of the mountains into the ice and snow of Baffin Bay. Everything seemed to be going to plan when suddenly we ran into our firsthand also our last, mechanical snag; part of the electrical system broke down. It could not be repaired in the air and so we were forced to turn and fly to the nearest airfield. This was Goose, a mere matter of twelve or thirteen hundred miles. We landed there after a ten hours' flight.[9]

'The trouble was soon put right and the next day we set out again for the North Magnetic Pole. It was essential that we waste no time, because in a few days' time the moon would sink below the horizon and we needed both the sun and the moon to fix our position. Once more we flew almost entirely either in or over cloud, with only a few glimpses of the ground. Seventy miles from where the Magnetic Pole was supposed to be, the magnetic compasses were still working normally and even when we arrived at the charted position, they still pointed on to the north-west. The Astronomer Royal had warned us that we might expect this; in fact, by a complicated mathematical analysis he had placed the Pole about three or four hundred miles farther ahead. So we flew on for about an hour, the compasses gradually becoming more and more erratic. Finally we turned and set course for Montreal, landing at Dorval airport after nineteen hours in the air.

'The next few days were spent visiting RCAF Headquarters and the Head Office of the Meteorological Services and in travelling across Canada. Everybody was extremely kind and helpful. Eventually on 24 May, we reached our final port of call, Whitehorse, in the Yukon. It is a beautiful spot, set deep amongst the Rocky Mountains, with the snow-covered peaks all round, the little town with its shacks and its stores somehow still smelling of the gold rush.

'Next morning we set out on our journey home, in which we hoped to fly past the northern side of the Magnetic Pole and so set some limits to its position. We flew out over the foothills of the Rocky Mountains, which gradually gave way to a flat, snow-covered plain. We came to the McClure Strait, named after some brigand ancestor of our wing commander observer and crossed the islands far to the north-west of Hudson Bay, passing within 600 miles of the North Pole itself. Here again our compasses went haywire, but it seemed to us that on the whole they tended to point southwards in the direction which the Astronomer Royal had predicted.

---

9. After taking off from Shawbury on 10 May 17,000 miles were flown in the Arctic regions alone. On return, *Aries* set out for Canada on 7 June and flew non-stop to the Canadian Central Navigation School at Rivers and broke the east-west crossing record. Later the England-to-the-Cape, London-Karachi and London-Darwin records were broken. *Lancaster – The Story of a Famous Bomber* by Bruce Robertson (Harleyford Publications Ltd 1964).

'Gradually the cloud below thickened and we caught only a few glimpses of the frozen sea and the black mountains of the islands rising out of the snow and ice. Just after we crossed the Greenland coast, however, the weather cleared and we flew for several hours over the great white ice-cap with its curious wind ridges known as Sastrugi. It was in the middle of this colossal sheet of snow, hundreds of miles from the mountains of the edge that an explorer friend of mine came upon the traces of a polar bear, travelling northwards!

'As we came near to the eastern coast, we saw great peaks jutting out of the snow to the north and as we passed by we saw ahead more great pyramids of black rock, the mountains of Scoresby Sound. Gradually the snow dipped away, pouring in vast drifts down between the jagged peaks, piling up in huge glaciers which rolled down through the mountains, cutting tremendous gorges through the black rock and sweeping majestically down to the frozen fiords.

'Perhaps this last sight of Greenland was the most beautiful of the whole trip. And yet the first glimpse a few hours later of the gentle Scottish Highlands was very good. Finally, [at 12:45 pm on 26 May] after 18½ hours' flying the skipper made a perfect landing on our own home airfield, Shawbury, having flown 110 hours in 16 days and covering 22,400 miles, roughly half of which lay within the Arctic Circle. [The crew had accumulated a massive quantity of information – 30,000 observations on magnetic phenomena alone.]. As we climbed out of the aircraft with the crowd milling around us some dreadfully perverse voice inside me kept repeating the words of a hymn that we used to sing at school which went something like – *'From Greenland's icy mountains; From India's coral strand'*. And so I could not help thinking of the war in Burma. I knew that the boys out there would be pleased that we had pulled it off. But also I realized that every day they were carrying out sorties which for courage and endurance left our polar trips a long way behind.

'And the results of the trip?[10] Well, we proved that an ordinary Royal Air Force aircraft with the normal equipment could fly in Polar Regions just as it could fly in any other part of the world. We brought back a

---

10. McKinley was awarded a Bar to his Air Force Cross. For their part in the mission, Anderson, McClure and Winfield were awarded the AFC, Wiggins and Dean the Air Force Medal, and the remaining crew members were commended for Valuable Services in the Air.

certain amount of data about the Magnetic Pole which caused a little stir in the Press, particularly in Canada. For some member of the crew had rashly guessed the position of the Magnetic Pole as being near the Sverdrup Island, which a Canadian reporter assumed to be in Russia! Consequently, Mr. Mackenzie King, the Prime Minister, in his election campaign that followed soon after, felt impelled to promise the electors with mock seriousness that if they would return him to power he would 'get the Pole back from the Russians!'

'Mary told Andy Mark III that I had been to the North Pole. Mark III was visibly impressed. 'Did Daddy hang his hat on it?'[11]

Postscript

In June 1945 a trip to Canada was made by *Aries* with Squadron Leader Imrie at the controls again. In September Wing Commander Clift took *Aries* on a tour of the Mediterranean; these two trips adding 17,300 miles in the log book. The next major flight was from RAF Thorney Island to South Africa in January 1946, when, during a normal liaison trip, *Aries*, piloted by Wing Commander Dunnicliffe made the first non-stop flight from Cairo to Cape Town and broke the England-Cape record with a time of 32 hours 11 minutes. In May *Aries* made a trip to Newfoundland, but the 3,950 mile round trip was considered a "local" flight by the E.A.N.S! On 20 August the famous Lancastrian departed for Australia and New Zealand via India and Ceylon and returned via Fiji, Singapore, Aden and Malta, breaking the records for London to Karachi (19 hours 21 minutes), London to Darwin (45 hours 35 minutes) and London to Wellington (59 hours 51 minutes) while clocking up 32,094 air miles.[12] In October-November 1946, Wing Commander Dunnicliffe flew *Aries* on its final liaison tour with a 12,309 mile round trip to Canada and the United States, taking in Labrador, Ottawa, New York, Ohio, California, Alabama, Florida, Maryland and Washington D.C.

---

11. *Pathfinders* by Wing Commander Bill Anderson OBE DFC AFC.
12. On both the May and August trips *Aries* was piloted by Squadron Leader Aldridge.

Chapter 13

Epilogue: Two Sides to Every Story

Monday 6th September 1943, 1948 hours GMT (2148 Double British Summer (local) time): A green Aldis light shone from the runway controller's caravan beside the runway threshold at RAF Binbrook, on the Lincolnshire Wolds, signalling clearance to take off. The concrete was damp from a recent shower of rain as, with its four Packard Merlin engines roaring under the strain of full power, a Lancaster surged down the runway and its pilot, Australian Flight Sergeant 'Jerry' Bateman, heaved his heavily-laden charge off the runway and into the air.

The sun had just set as Lancaster B III W5005 AR-L, climbing slowly, thundered off into the gathering darkness. This was to be the 30th operation over enemy territory for this particular Lancaster. It was the crew's 12th 'op' of their 30-op tour with on 460 Squadron RAAF *and the 10th in 'their' aircraft, which was named 'LEADER' and which was adorned with some special nose art.*

Squadron Leader Clive Rowley MBE RAF (RETD)

There have been, according to various tomes, 36 Lancasters that famously recorded 100 operations or more during their wartime careers. To date three wartime centurions have been represented by the Battle of Britain Memorial Flight (BBMF) Lancaster PA474 'City of Lincoln' In 2007 the BBMF Lancaster emerged from its winter major servicing wearing the dual markings of 100 and 550 Squadrons' Lancaster III EE139 *Phantom of the Ruhr,* the port side being painted as it would have been at the end of its service life on 100 Squadron with 'HW-R' on the fuselage and with 'BQ-B', the letters of 550 Squadron added on the starboard side.

When in 2017 the BBMF Lancaster appeared wearing a new WW2 colour scheme, the code letters 'VN-T' for 'Tommy' were applied to the starboard side to represent LL922 on 50 Squadron and honouring the crew captained by Warrant Officer 'Dougy' Millikin, whose nephew,

'Millie' Millikin, was at the time the OC BBMF. 'T for Tommy' became their allocated aircraft for their tour. At the age of 28 'Dougy' Millikin was older than many Lancaster pilots and captains. He was now an experienced pilot with over 1,000 flying hours, having flown twin-engine Bristol Blenheims on 30 Squadron at RAF Habbaniya, Iraq in 1938. His wireless operator was 21-year-old Sergeant John (Harry) Tait (nicknamed after the comedian Harry Tate). John had moved to Irby, a village on the Wirral, Merseyside, with his family when he was five years old. His father, who was originally from Hamilton, Scotland had emigrated to the USA in the 1920s where he farmed in Montana. In 1928 he decided to bring the family back to England and John grew up on the Wirral. When war broke out in 1939 John was only 16 and had to wait until 1941 before he was old enough to sign up for military service. He chose the RAF because 'I knew a couple of local lads who were already fighting for Britain at this time and I really fancied the opportunity to be in the RAF as opposed to other services.' After joining the RAF and spending six weeks 'square bashing' training, John Tait volunteered as a WOp/AG and was posted to Blackpool and then onto Yatesbury, in Wiltshire, where he did his radio training, by the end of which he could average 22 Morse code words per minute. He then had to do another eight-week course at Stormy Down for his air gunner training. This was the first time that he met Warrant Officer 'Dougy' Millikin, who would be his skipper for the next two years. 'In the air he was fantastic, on the ground he was good fun. Once he got in the aircraft everything was according to the book. He would order the crew onto oxygen at 2,000 ft to ensure that no one forgot when we got to 10,000. '

The fledgling bomber crew began forming on 29 OTU at RAF Bruntingthorpe in Leicestershire where they flew the Vickers Wellington. Flight Sergeant 'Jimmy' James the 30-year-old flight engineer had been a ground engineer in South Africa who had volunteered for aircrew to get back to the UK and who 'really fancied the opportunity to be in the RAF as opposed to other services'. Pilot Officer 'Wally' Pugh the 21-year-old year bomb aimer and a professional pianist, was from Liverpool. Pilot Officer 'Jimmy' Marlow the navigator, also 21, had worked for the Air Ministry before the war and was a civil servant. He had wanted to get a commission which he achieved, Sergeant John 'Jock' Brockett, the Glaswegian mid-upper gunner was a 'real rough diamond' when, if there was any trouble at all, 'Jock' sorted it out.' He and Sergeant Johnny Austin, the rear gunner, who subsequently got frostbite and was de-mobbed, were both only 18 years old at the start of their tour of bomber operations. We were all posted to Swinderby where we gained our first experience of flying as a seven-man crew. We initially trained on Short Stirlings at Scampton, which had four engines and which we thought were pretty monstrous, as we went about learning our responsibilities in preparation for flying on 'ops'. After a couple of months we were fortunately moved off Stirlings and began our conversion training at Syerston on that wonderful aircraft, the Lancaster bomber.'

When the crew had completed their OTU course and Heavy Conversion Unit training they were posted at the beginning of April 1944 to join 50 Squadron at RAF Skellingthorpe. Over the next three months the Millikin crew flew on operational bombing operations over Nazi occupied Europe. A few were flown into Germany, such as their sixth 'op' to Munich on 24 April 1944, a trip lasting over ten hours when Doug's wife went into labour and their firstborn child was born. On the way back at 0600 John Tait tuned into the BBC. The broadcast said 'last night our bombers went to Munich…. and John said, 'We hadn't got back!' John Tait said. 'I knew that I was lucky to be part of a fantastic crew, who did our bit fighting over France and Germany and I remember the great friendships we formed when we were back in Lincoln. 'Dougy' had a black donkey he took on every trip and I had a doll I carried on every trip. I do remember the time we were caught in all those searchlights on the raid to Schweinfurt [26 April 1944]. When a searchlight picks you up, it's like being in broad daylight. You could see everything inside the aircraft. You had to corkscrew to get out of it. Well, 'Dougy' was very good at that! The routine was diving, rolling, climbing, rolling, diving, rolling and my logbook was floating in front of me!' On 8/9 May they dropped a 4,000 pounder on the airfield and seaplane base at Lanveoc-Pulmic at Brest. Because it was so close it was 'just a local trip' according to John Tait. On 10/11 May they were one of the Lancaster crews that bombed the marshalling yards at Lille from 8,250 ft! On 18 May they went to Paris (Juvisy) to destroy the works near the rail yard and attacked from 10,500 ft! On 21 May it was Duisburg. 'The op report said 'fighter activity between target and coast was very considerable' recalled John Tait 'but, it was just one of those things!'

On D-Day itself, 6 June 1944, 'Dougie Millikin's crew flew two relatively short trips to bomb German coastal artillery batteries at 0500 hours, immediately prior to the first invasion landings and then another sortie that night to bomb crossroads at Argentan to restrict the movement of German reinforcements towards the beachheads. On their 28th 'op' to bomb a target in France, the crew's Lancaster was attacked by German night fighters three times, but they managed to evade the attacks and returned safely to base. On 29 June they were one of 286 Lancaster crews that attacked two flying-bomb launch sites in the Pas de Calais. 'It was a daylight mission' recalled John Tait. 'I remember daylight missions. I tell you what, there were bombs dropping all around 'cos the Halifaxes were higher than us, the Stirlings were down below and it was mayhem. We'd watch the bombs land.' On 4 July near Paris they were attacked three times by night fighters but came home safely once again. By late July 1944 'Dougy' Millikin and his crew had completed more than their required number of operations for a full tour. 'Dougy' had flown 35 ops and John, 34. Twenty-seven of these were flown on LL922.

'T-Tommy' was subsequently lost on the raid on German tank emplacements in the woods near Secqueville, France on the night of Monday, 7/Tuesday, 8 August 1944, just twelve days after the Millikin crew last flew it. En route to the target an order was received by with 26-year-old Flight Lieutenant Richard Silvio Palandri's crew to abort the operation as the Americans had captured the area concerned. LL922 however, was almost6 immediately attacked by a German night fighter and went into a tail spin on fire but Palandri managed to straighten out the aircraft which allowed four of the crew time to bail out. Sergeant William 'Geordie' Johnson the 21-year-old mid-upper gunner, claimed to have shot down the twin-engined night which followed the bomber down before it crashed at Frénaye, 3 kilometres ENE of Lillebonne with the bomb load still on board. Johnson had to wrench the cords of his parachute because they were so low to the ground and the explosion blew him upwards. Two French schoolboys who had gone to take a look were killed in the explosion.

Johnson was picked up by the Maquis and went on several raids taking part in sabotaging and blowing up telegraph cables and ammo dumps. Palandri, Flight Sergeant Reginald John Owen the 21-year-old navigator and Sergeant Arthur Donald Mellish the 22-year-old WOp/AG were killed. Sergeant Johnny B. Firth, the 19-year-old flight engineer became a prisoner of war. The air bomber, Flight Lieutenant Edward H. E. Hearn DFC, who had taken the place of Flight Lieutenant Mick Manus, the usual bomb aimer who was unable to fly on this operation due to sickness, and Sergeant William Johnson and Flight Sergeant Arthur Robson Meredith the rear gunner, evaded capture. This was the crew's 29th op; eleven on which had been flown on LL922. It is thought that the Luftwaffe pilot that shot them down was either Hauptmann Heinz Rökker or Leutnant Johannes Strassner both of II/NJG2. Palandri left a widow, Kate from Miriam, Italy, a Night Nursing Attendant living in Calne, Wiltshire and three children.

'Dougy' Millikin had by now been commissioned and held the rank of Flying Officer. He was subsequently awarded the DFC (gazetted in November 1944) and he went on to fly a second tour of 'ops' on 150 PFF Squadron.

John Tait went on to 35 Squadron on Pathfinders on his second tour, remaining in the RAF for a while after the war and completing a tour of America with a display team of twelve Lancasters. When he left the RAF, as a Warrant Officer, he went home to Irby and to his wife, Beryl, who he had met at a dance during the heydays of the war, to get a job and start a family. He spent most of his working life as a building surveyor and then in his fifties opened several health shops. Passionate about sport, John was a prolific goal scorer for Heswall and North Newton Football Clubs, regularly scoring in excess of 30–40 goals per season. John also became president of the Irby Club where he used to enjoy a pint after returning from duty during his war days. He had now been a member of the club for 75 years. In April 2017 John was 94 years old.

In 2017 when the BBMF Lancaster's starboard side began carrying the markings of 'T-Tommy' on 50 Squadron the port side received the letters 'AR-L' and *'LEADER'* replicating Lancaster W5005 that served on 460 Squadron RAAF at Binbrook, nine miles SW of Grimsby, where the Australian squadron was based from 14 May 1943 to 28 July 1945. In July 1943 when the Lancaster had already completed twelve 'ops' on the Australian squadron, it was the crew skippered by Scottish pilot, Sergeant (later Flight Sergeant) J. D. 'Jock' Ogilvie, one of the few British RAF pilots on this Australian squadron, who came up with the idea of the nose art on W5005. Because half the crew were Scottish and half were

Australian personnel AR-L carried the famous nose art consisting of a kangaroo wearing flying boots and playing lustily on the tartan bagpipes and it was Flying Officer Thomas Victor 'Vic' Watts who created it. Having arrived at Binbrook in May 1943 he was to serve on 460 Squadron as a navigator for two tours of operations, the first ending in November 1943 and the second from October 1944 to May 1945, by which time he was a Flight Lieutenant with two DFCs. Watts painted much of the nose art on 460 Squadron's Lancasters, with style and humour. The Ogilvie crew flew W5005 on nine 'ops' between 21 June and 29 July 1943, three of which were aborted for a variety of reasons.

By the spring of 1944 W5005 had by now been re-coded as AR-E^2 and its last 'op' on 460 Squadron, its 51st, was to Karlsruhe on 24/25 April. On landing from this sortie the aircraft did not touch down until well into the available runway length. In an attempt to avoid going off the end of the runway, Australian Flight Sergeant (later Flying Officer DFC) 'Dan' Cullen tried to turn the aircraft at speed, but this resulted in the starboard undercarriage collapsing. However, the Lancaster was quickly repaired and then transferred to 550 Squadron at North Killingholme where it was re-coded as 'BQ-N' and known as 'N-Nan', although it retained its kangaroo and bagpipe nose art, flying its first 'op' with its new unit on 9/10 May 1944 against coastal batteries at Mardyck, near Dunkirk. W5005 was locally modified by 550 Squadron with an 'under gun' to counter 'Schräge Musik' night fighter attacks from below. A hole was cut in its floor behind the bomb bay and a 0.5 heavy machine gun was mounted on a rail pointing down and rearwards. A deflection plate was fitted to shield the gunner from the slipstream. The aircraft flew a number of operations with an eighth crew member – the mid-under gunner – but his field of view was very limited and it proved to be ineffective.

When, in 2017 it was decided to paint the port side of the BBMF Lancaster in the famous markings of 'AR-L *LEADER*' a potential problem arose. Unlike the straightforward application of VN-T's details on the starboard fuselage, there remained some doubt as to the correct serial number and code letters of W5005 on the port side. Published transcripts seemed to indicate that the wartime codes for the aircraft were that of 'N for Nancy' flown by Pilot Officer 'Stan' Ireland. Squadron Leader Clive Rowley MBE, a previous OC BBMF who took on the investigation to ensure that the correct codes and serial number were used, explains. 'The mix-up over the 'Stan' Ireland crew and which Lancaster was theirs – JB607 'AR-N' or W5005 'AR-L' – did take some sorting out. It was the fact that JB607 was shot down on its ninth op on 29 December 1943 that convinced me of the widely reported error and that JB607 could not have been *LEADER*, which had thirty ops on the bomb log in photos taken on 9 September 1943. There was a press day at Binbrook that day and W5005 'AR-L *LEADER*' featured in several photos, one with the Ireland crew standing under the nose (only six of the seven men are visible) so some people jumped to the conclusion that this was the aircraft in which they were shot down in on 29 December 1943 – quite a 'leap'! Incidentally the photo taken on 18 August 1943 of the 'L-Leader' kangaroo nose art with 24 ops on the bomb log and a hat-wearing aircrew chap in the cockpit is of Flight Lieutenant A. S. MacWilliam DFC, the 460 Squadron Gunnery Leader and not 'Stan' Ireland. Once I had waded through the 460 Squadron ORB pages and counted up the ops by JB607 (nine total) I dropped my interest in that aircraft.'[1]

W5005 flew 43 'ops' on 550 Squadron until its luck ran out on 26/27 August 1944 on returning from its 94th operation of the war to Kiel with flak damage. Flight Sergeant R. 'Hoppy' Hopman RAAF was attempting a flapless landing at North Killingholme at 0150 hours, but he misjudged the approach and ditched in shallow water on mudflats in the Humber Estuary near Killingholme Haven. The crew of five Australian and two British crewmen were uninjured and paddled ashore in their dinghy. But while some did not mourn W5005's passing – Wing Commander Jack Harris on 550 Squadron said, 'There were sighs of relief when 'N-Nan' ditched in the Humber; the aircraft flew very badly', W5005 had narrowly missed out on becoming a 'ton up', 'Centurion' Lancaster.

Today there are seventeen surviving Lancasters in the world but only the Battle of Britain Memorial Flight Lancaster and one in Canada operate in flying condition.

---

1. *Royal Air Force Memorial Flight Club Official Yearbook 2017.*

Index

Allwright, Flight Sergeant Ernest Frank 70
Anderson, Flight Lieutenant John Thomas 93, 98
Anderson, Wing Commander Bill OBE DFC 228–229, 231–241
Annan, Wing Commander Bob 186
Arents, Will 102
Athey, Sergeant Stanley Tallintire 99
Atlee, Clement 225
Augsburg 1–23, 177

Backwell-Smith, Squadron Leader Raymond 107, 109–110
Bader, Douglas 226
Bardney 106
Barnes, Pilot Officer Leonard Alfred 'Barney' 119–122
Battle of Britain Memorial Flight (BBMF) 242
Baxter, Wing Commander Ronald Edward 31, 34
Bazin, Wing Commander James Michael DFC 216, 222
Beck Row 143
Beck, 'Pip' 64
Beckett, Warrant Officer John Frank 'Joe' DFM RAFVR 9, 13–14
Berlin airlift 230
Berlin 74, 88–89, 176–177, 202
Berrington, Flying Officer John Raynor 106
Berry, Flight Lieutenant Leslie Frederick 'Peter' MiD DFC 154
Berry, Flight Lieutenant Surender Lal 'Sammy' DFC RAFVR 127–128, 144, 148–152
Biennials 185
Binbrook 247
'Bird In Hand' 143
Birkett, Sergeant Nicol Thomson 14
Birnie, Pilot Officer John 'Jock' 179
Blackbushe 174
Blanchard, Flight Sergeant Eric DFC 63, 72–73, 90

Blott, Flight Lieutenant Walter 'Bill' RAFVR 112
Bochum 65, 70
Booth, Flight Sergeant Mike 87
Bourn 98
Branch, Flight Sergeant Albert Charles DFM 94–95
Breighton 118
Bremerhaven 90
Britton, Pilot Officer 96
Brown, Sergeant K. D. 71, 73–74
Brunswick 49–50
Bruntingthorpe 58–59, 243
Buckham, Flight Lieutenant Bruce DFC RAAF 211, 224
Buggy, Pilot Officer Norman 77
Bulpett, Squadron Leader Ed 102
Burgwal, Rudolf 'Rudy' Franz 66
Burpee, Flight Sergeant Lewis J. DFM RCAF 32
Byams, Guy 211

Calvert, Leslie 'Cal', T 24, 26–28, 32–34, 36–41
Cameron, Pilot Officer William Parmenas 99
Campbell, Sergeant 'Pete' 191, 196, 203, 207, 209
Cap Gris Nez 181
Carnaby 57
Castle Donington 48
Castrop-Rauxel 199
Chadwick Bates DFC, Pilot Officer Arthur George Jackson x–xvii
Chambry 180
Chedburgh 191
Cheshire, Wing Commander Leonard 216
Chevreuse 182
Churchill, Winston 23
Clegg, Flying Officer James GM RAAF 116
Clements, Flying Officer Robert Sinclair RCAF 77
Cochrane, Air Vice Marshal Ralph 95

Cochrane, Squadron Leader Archibald George Alexander DFC 94
Collier, Wing Commander John David Drought 'Joe' DFC 5–6
Cologne 71, 196–199
Colville, Flying Officer Alexander Colborne RCAF 114
Coningsby 102–103
Cooke, Alfred Alistair KBE 63
Courtrai 181
Craig, Flight Lieutenant James Fraser DFC RNZAF 116
Crum, Warrant Officer Herbert V. DFM 9, 13–15

D'Ombrain, Pilot Officer Peter Charles Lewis RAAF 154–156, 159–160, 163, 167
Davidstow Moor 76
de Jongh, Elly 66
de Waard, Harry 92, 101
Devas, Wing Commander Guy 186
Deverill, Flying Officer Ernie 4–5, 17, 20–21
Dimbleby, Richard 226
Dorehill, Pilot Officer Patrick Arthur 7, 11–12, 15–16, 22
Dortmund 34, 69, 181, 199, 201
Dortmund-Ems Canal 5
Dowty, Sergeant Bertram Arthur 13–15
Dreux 183
Drummond, Flight Sergeant Kenneth RNZAF 113
Duisburg 55, 181, 194–195
Dunham, Wing Commander Peter, Francis DFC 208–210
Dunholme Lodge 75, 106
Düsseldorf 70, 77, 195

Eames, Sergeant Roy Clive 144–145, 148
Earnshaw, Sergeant Frank 95
East Kirkby 60, 77, 98, 112, 119
Eddy, Flight Lieutenant Charlie RNZAF 179
Edgehill 191
Elliott, Flying Officer James McPhail 77
Erkenshwick 209
Essen 56, 180, 196
Evans, Albert Edward 95

Farnborough, RAF 233
Feltwell 191
Finningley 8
Firth, Pilot Officer Christopher 99
Fiskerton 117
Forster, Flight Lieutenant 96

Foster, Clarence William 95
Friedrichshafen 143–144
Fynn, Flight Lieutenant Terence Hugh 106, 119

Ganderton, Flying Officer Arthur 117
Gardner, Pilot Officer Alfred Edward Walter 90
Garwell, Flying Officer John 'Ginger' DFM 9, 12, 16–17
Gearing, Sergeant Leonard 'Len' Thomas 156, 160
Gelsenkirchen 74
Gibson, Wing Commander Guy DSO DFC* 32, 69
Ginn, Sergeant 'Jimmy' 191, 197, 206
Glendenning, Raymond 226
Gobbie, Flight Lieutenant Anthony Francis DFC 90
Gomorrah, Operation 77–78
Gordon-Watkins, Wing Commander William David DSO DFC DFM 155
Gransden Lodge 99, 114
Grantham 3
Graveley 117
Gregory, Flight Lieutenant Victor 204–205
Greisert, Hauptmann Karl-Heinz 'Heine' 13

Hadder, Sergeant Leslie Arnold 156, 160
Halley, Flight Lieutenant 180
Hallows, Flying Officer Brian Roger Wakefield 'Darkie' 7–10, 18, 21
Hamburg 77–78, 87–88
Hannent, Alice 192
Hannover 89
Haralambides, Flight Sergeant Stephanos 189
Harris, Air Marshal Sir Arthur T. 'Bomber' 6, 23, 77, 186
Harris, Wing Commander Jack 248
Hartford Bridge 174
Hawkins, Flight Lieutenant Jack BBMF 103
Haye, Pilot Officer Jan Bernard Marinus 'Dutchy' 65–66, 77
Hayes, Sergeant Joe 179
Hayes, Sergeant Ron 145–146
Hemswell 42
Heston 75
Higgins, Flying Officer Robert Josephus Constable 198
Holmes, Flying Officer Roland 55
Honington 228

Hopman, Flight Sergeant R. 'Hoppy'
  RAAF 248
Howe, Pilot Officer Joseph 60, 90
Howes, Flight Sergeant E. J. 63
Hurricane', Operation 56, 195
Hutchins, Sergeant Glyn Davies 99

Ifould, Pilot Officer Edward Lister
  'Ding' 10–11, 19–20
Irons, Ronald Percy Irons DFM 20

Jamieson, Flight Sergeant Laurence
  Seymour 'Pop' 161–162, 165
Johnson, Warrant Officer 1 O. O. 100–101
Jones, Flying Officer William E.
  'Jonah' 105
Josling, Sergeant John Basil 70

Kaiser, John William 'Jack' 190–210
Karlsruhe 180
Kelstern 99
Keon, Flying Officer Howard 'Howie' 191,
  198–199, 203, 208–210,
Kirmington 99, 105, 124
Kleve 194
Klufas, Squadron Leader RCAF 174, 188
Köberich, Oberleutnant Günther 117
Koblenz 205
Koch, Unteroffizier Herbert 116
Koss, 'Chuck' 176, 180
Krefeld 71

L'Hey 185
La Spezia 33
Lamblin, Sergeant 71
Laon 178
Lashford, Flight Sergeant F. W. RAFVR
  DFC 98
Lau, Oberleutnant Fritz 118
Le Havre 184
Le Mans 181
Learoyd, Wing Commander Roderick
  Alastair Brook 'Babe' VC 4–5
Leatherdale, Squadron Leader Francis
  'Frank' Ridley DFC 170–189
Leatherhead 174
Leipzig 89–90, 92–93, 95–96, 99
Leitch, Flying Officer James Westwood
  DFC RAAF 98
Leuchs, Major Rolf 116
Linton-on-Ouse 113, 115
Lisieux 182
Little Snoring 93, 98
Little, Flying Officer Ralph Robert
  RCAF 58

Long, Flight Sergeant Arthur 'Steve'
  Stephen 162–163, 166, 168
Lossiemouth 215
Ludford Magna 42–43, 49, 57
Lumgair, Flight Sergeant Norman Andrew
  RCAF 113
Lundy, Flight Lieutenant Dennis James
  DFC 226
Lütje, Hauptmann Herbert Heinrich
  Otto 66–67

Mackie, Squadron Leader George 'Mac'
  DSO DFC 170, 172–174, 177–178, 184,
  187–188
MacKinnon, Flight Sergeant Murdock
  Daniel RCAF 146
MacWilliam, Flight Lieutenant A. S.
  DFC 248
Mannesmann tubular steel works 77
Mannheim 89
Manston 28
Manton, Squadron Leader Richard
  John 93–94, 101–102
Martin, Sergeant James George Louis 63,
  70, 90
Mattock, Sergeant G. R. 112
Mauthausen 66
McClure, Kenneth Cecil 234–235
McKechnie, Pilot Officer Donald S.
  RCAF 170
McKee, Flight Sergeant W. S. J. RCAF 145
McKinley, Wing Commander 230,
  233–234
McLeish, James Campbell DFC 118
McMullen, Group Captain Colin
  Campbell AFC 216, 222
Meggason, Flying Officer Ross 191,
  194–195, 198, 200–203, 205, 207–208,
  210
Mepal 178
Methwold 203
Meyer, Flight Lieutenant William
  Alexander DFC 118
Milan 88
Mildenhall 143, 153–154, 156, 162, 169,
  205
Miller, Glenn 204
Millikin, 'Millic' 243
Millikin, Warrant Officer 'Dougy' 2
  42–246
Mönchengladbach 77
Montdidier 184
Morley, Jack 'Sheff' 42–59
Müller, Unteroffizier 'Hans' 119
Munich 89

Murphy, Sergeant D. 112
Mycock, Warrant Officer Thomas James DFC 3–4, 9

Nantes 183–184
Nettleton, Squadron Leader John Dering VC 6, 9–11, 18, 22, 75
Neuss 196, 208
Newcomb, Flight Sergeant Richard William 89–90
Newmarket racecourse 96
Nonnenmacher, Unteroffizier Emil 114
Norris, Sergeant Raymond Geoffrey 'Geoff' 156–159, 161–163, 165–168
North Killingholme 116, 247–248
Northrop, Wing Commander Joe 105, 110–112
Norwich 64, 90–91, 156, 192
Nuffield Trust 91
Nuremberg 143
Nystrom, Flight Sergeant Stanley Arthur 'Sam' RAAF 160, 162

Oakington 98, 117, 188
Oberhausen 70
Oboe 77–78, 185
Oesau, Major Walter 'Gulle' Oesau 12
Ogilvie, Flight Sergeant J. D. 'Jock' 246

Painter, Pilot Officer Kenneth 98
Parker, Pilot Officer Frank Ernest Saville 98
Peall, Anthony Paul ('Buster') 1, 12
Penkuri, Flying Officer Kaiho 'Tommy' RCAF 113
Penman, 'Jock' 4, 9, 18–20, 22
Pickett, Flight Sergeant Johnny RNZAF 76
Pilsen Škoda armaments factory 66
Pohl, Unteroffizier Walther Pohl 13

Randall, Flight Lieutenant Francis Reginald 'Frank' 153
Ransom, Flying Officer William H. RCAF 144
Ravensbrück 66
Reid, Flight Sergeant Frank Bruce 164
Rhodes, Sergeant George Thomas 'Dusty' Rhodes 9
Richer, Leo 210
Robbins, Edward A. 'Ted' 24–29, 34–40
Rodbourn, Pilot Officer Kenneth 121
Rodley, Flying Officer Eric E. 'Rod' 5–6, 9–10, 17, 18–19
Rökker, Oberleutnant Heinz 108
Rowland, Ann 103

Rowland, Peter 102
Rowley, Squadron Leader Clive MBE RAF (retd) 242, 247–248
Rupp, Unteroffizier Bruno 115

Saarbrücken 194
Sainte-Hippolyte 144
Sandford, Flight Lieutenant Reginald Robert 'Nicky' DFC 9, 12
Sandtoft 42
Scampton 5, 65–66, 68–69
Schnaufer, Heinz-Wolfgang 98
Schön, Leutnant Walter 36
Schräge Musik' 182
Scott, Squadron Leader A. R. DFC* RNZAF 193, 200, 210
Seager, Flight Sergeant Dennis Ernest Arthur 76
Searby, Wing Commander John 31
Seddon, Flight Lieutenant 180
Shaw, Pilot Officer Fred 204–205
Sherwood, Squadron Leader John Seymour DFC* 3, 9–10, 18–19
Shorthouse, Flight Lieutenant John Sidney DFC 75–76
Siegen 204–205
Simard, Pilot Officer Arthur Gerald Sylvain 115
Simpson, Pilot Officer J. S. 96
Sinding, Christian 94
Sirman, Sergeant Norman 107
Skellingthorpe 244
Smith, John George Smith 90
Smith, Sergeant John George 60
Smith, Squadron Leader R. H. 188
Smith, Squadron Leader Ronald Edward Sidney DFC 65
Söthe, Hauptmann Fritz 153–155, 167
Sparks, Harry 96–98
Sparks, Squadron Leader Ernest Neville Monkhouse 105
Squires Gate 22
Stettin 51
Stevens, 'Steve' 60–91
Stevens, Maureen (nee Miller) 64, 68–69, 76, 90–91
Stuttgart 56, 104–126, 196
Swales, Wing Commander Ian Clifford Kirby DFC DFM 148
Swinderby 244

Tait, Wing Commander James Brian 'Willie' DSO DFC 216, 222–224
Taylor, Flight Sergeant Roy Leonard 95
Telford, Sergeant William Clare RCAF 192

Tempsford 182
Thomas, William Kenneth DFC 152
Thompson, Walter R. DFC* 94
Thorney Island 241
Tirpitz 211–227
Trappes 154, 160
Trier 205
Tuddenham 190, 192, 194, 199, 201, 203
Turin 74–76
Twinwood airfield 204

Vaires 188
Venter, Sergeant Peter Johannes 12
Villacoublay 205
Villeneuve-Sainte-Georges 178
Villers-Bocage 186–188
Vinke, Oberfeldwebel Heinz 101
Vohwinkel 207
von Bonin, Hauptmann Eckart-Wilhelm 112–113

Waddington 2, 4, 119, 231
Wagner, John 175–176, 179
Walcheren Island 193–194
Wallis, Dr. Barnes 214
Warboys 93, 101,189
Ward-Hunt, Squadron Leader Peter 31
Waterbeach 89, 182, 203
Watson, Flight Lieutenant James 'Jimmy' Andrew RCAF 127, 144, 146–150
Watson, Flight Sergeant Donald Moulton 98

Watts, Flying Officer Thomas Victor 'Vic' 247
Watts, Pilot Officer Robert Hamilton RAAF 101
Waugh, Flying Officer Robert John 90
Weighell, Sergeant George Myron 99
West, 1st Lieutenant Don 77
West, W. E. 211
Westcott, OTU 155, 162
Wigsley 45, 63, 68
Wild Boar' 183
Wilhelmshaven 56
Willcocks, Sergeant Edgar Harold RAFVR 99
Williams, Flight Lieutenant Reg 210
'Window' 78, 87
Winfield, Wing Commander Dr. Roland DFC 223–224
Witchford 116, 170–172, 174–175, 177, 179–180, 203
Wood, Pilot Officer Kemble Russell RAAF 99–100
Woodbridge 107
Woodhall Spa 2–3, 9–10, 99
Wooldridge, Flight Sergeant Arthur 'Bill' 190, 192, 203, 206, 208–210
Wooldridge, Michael 192, 208–210
Wratting Common 203
Wynyard, Flight Lieutenant Tony 74
Wyton 93–94, 105

Yates, Flight Sergeant Wilson 89